CURRENT STEM

CURRENT STEM

VOLUME 1

CURRENT STEM

MAURICE HT LING (SERIES EDITOR)
PRINCIPAL PARTNER, COLOSSUS TECHNOLOGIES LLP, SINGAPORE

Current STEM. Volume 1
Maurice HT Ling (Editor)
2018. ISBN: 978-1-53613-416-2
(hardcover)
2018. ISBN: 978-1-53613-417-9
(eBook)

CURRENT STEM

VOLUME 1

MAURICE HT LING
EDITOR

nova
science publishers
New York

Library of Congress Cataloging-in-Publication Data

ISBN: 978-1-53613-416-2

Published by Nova Science Publishers, Inc. † New York

CONTENTS

Preface		**vii**
Chapter 1	Digital Organism Simulation Environment (DOSE) Version 1.0.4 *Clarence FG Castillo and Maurice HT Ling*	**1**
Chapter 2	Socio-Cultural and Behavioral Factors Influencing Risk of HIV among Women in Grenada *Winta Ghidei and Shelly Rodrigo*	**107**
Chapter 3	A Cryptography Method Inspired by Jigsaw Puzzles *Maurice HT Ling*	**129**
Chapter 4	COPADS V: Lindenmayer System with Stochastic and Function-Based Rules *Maurice HT Ling*	**143**
Chapter 5	COPADS VI: Fixed Time-Step ODE Solvers with Mixed ODE and Non-ODE Function, and Script Generator *Maurice HT Ling*	**173**

About the Editor 213

Series Description 215

Index 219

PREFACE

Science, technology, engineering, and mathematics – collectively known as STEM – are rapidly advancing fields in their own rights. As a book series, *Current STEM* aims to be a friendly forum for both academic researchers and industrial practitioners to present their work as book chapters. Hence, the chapters should be varied, and this is intended. *Current STEM* encompasses the type of work to encourage a generation of researcher-practitioners.

The author hopes this inaugural volume will be a pleasure to a wide audience.

Maurice HT Ling, PhD

In: Current STEM. Volume 1
Editor: Maurice HT Ling
ISBN: 978-1-53613-416-2

Chapter 1

Digital Organism Simulation Environment (DOSE) Version 1.0.4

Clarence FG Castillo[1,2] and Maurice HT Ling[1,3,*]
[1]Colossus Technologies LLP, Singapore
[2]School of Information Technology, Republic Polytechnic, Singapore
[3]School of BioSciences, The University of Melbourne, Australia

Abstract

Evolution is a fundamental aspect of biology but examining evolution is difficult, time-consuming and costly. At the same time, molecular analysis of biological organisms is generally destructive, which presents a conundrum between observing possible evolutionary outcomes and in depth molecular analysis to decipher the corresponding evolutionary outcomes. Artificial life simulations via the use of digital organisms (DO) had been proposed as a feasible means of examining evolution *in silico* and had yield biologically relevant findings. Being digital, identical replicates can be made for analysis; thereby, resolving the conundrum. Recently, original implementation of DOSE (Ling, 2012a) had been improved (Castillo and Ling, 2014a) for use as a Python library for simplified construction of simulation, enabling database logging and revival of

* Corresponding Author Emails: mauriceling@acm.org, mauriceling@colossus-tech.com.

simulations. This manuscript documents the implementation and improvement of DOSE, which is released as DOSE version 1.0.4 (https://github.com/mauriceling/dose/releases/tag/v1.0.4) and licensed under GNU General Public License version 3. DOSE codebase is hosted and available for forking at https://github.com/mauriceling/dose.

Keywords: simulation environment, digital organism, artificial life

INTRODUCTION

Evolution is a fundamental aspect of biology where organism adapt to the environment, resulting in emergent responses. One of the earlier iconic examples of evolution (Clarke et al., 1985) is the phenotype colour change of peppered moth from light coloured to dark coloured, as observed by higher occurrence of dark coloured peppered moth, during increased air pollution in the second half of 1800s. This is followed by reversal from darker coloured to lighter coloured, where occurrence of lighter coloured pepper moth increased from 6% to 30%, during reduced air pollution from 1959 to 1984 (Clarke et al., 1985). More recent examples include *Anolis carolinensis*, a lizard, evolving larger toepads to inhibit higher perches after invasion by another lizard, *Anolis sagrei* (Stuart et al., 2014). House mouse (*Mus musculus domesticus*) developing resistance to warfarin (Song et al., 2011), an anti-coagulant used as rodenticide.

However, testing evolutionary hypotheses is a challenge (Batut et al., 2013) as it is highly time consuming and expensive, if not impossible. Generally, many generations are required for adaptation; hence, the primary hurdle of evolutionary experiment in biology is the generation time. Thus, most experimental evolution can only be carried out using organisms with short generation time (Buckling et al., 2009); such as bacteria, yeast, or viruses. In addition, most genetic analysis to find out the basis for such adaptations requires the extraction of DNA from the cells; hence, it is destructive testing. At the same time, it is generally prohibitively expensive to analyse the genome of each cell. As a result, most genetic analyses are performed on groups of cells rather than individual cell.

Christopher Langton had conceptualized that by casting chemical reactions, reactants, and products into computable operations, operands, and outputs respectively, it may be possible to simulate artificial life (organisms in the digital world, or digital organisms) as cellular automata "living" on artificial chemistries (Langton, 1986). A study on the definition of life has also concluded that artificial life can be considered "alive" (Koh and Ling, 2013). Being digital, an organism or a cell can be replicated, by making another computational instance of the cell or by "reading" the cell from hard disk without the need to destroy the original copy, for analysis. Thereby, removing the dilemma related to destructive testing – the very sample of cells that is being genetically analysed are the cells containing emergent features. Moreover, it is also possible to analyse each genome as computational analysis of a genome generally requires less cost compared to biological analysis. As a result, a large volume of studies had used digital organisms to study biological evolution (Ling, 2014); including the effects of genome size (Gupta et al., 2016), effects of population (LaBar and Adami, 2016), speciation (Anderson and Harmon, 2014), and resistance to quorum quenching (Beckmann et al., 2012). Batut et al. (2013) argue that artificial life or digital organisms (DO) is a valuable tool to enable experimental evolution despite its drawbacks as repeated simulations can be carried out with recording of all events. Bersini argued that artificial life and theoretical biology shared many common grounds and presented genetic algorithm (GA) as an important model to bridge the two fields (Bersini, 2009).

A Python implementation of GA, grounded on the biological hierarchy from genome to organisms to population (Lim et al., 2010), had been used as a basis to develop into a simulation framework for digital organisms, known as Digital Organisms Simulation Environment (DOSE) (Ling, 2012a).The genetic material of DOSE organisms is an executable DNA based on a parameterless 3-character instruction code, known as Ragaraja (Ling, 2012b), which Ling (2012a) argued that such an instruction set mimics the naturally occurring DNA.

Recently, the original implementation of DOSE (Ling, 2012a) had been improved (Castillo and Ling, 2014a) for use as a Python library for simplified construction of simulation, enabling database logging and revival of simulations. A utility case study of DOSE demonstrated that adjacent migration, such as foraging or nomadic behavior, increases heterozygosity (local genetic distance) while long distance migration, such as flight covering the entire ecosystem, does not increase heterozygosity. These results are consistent with previous studies (Raymond et al., 2013; Relethford, 1988). In addition, DOSE had been used to simulate antibiotics resistance (Castillo and Ling, 2014b; Castillo et al., 2015). This manuscript documents the implementation of DOSE version 1.0.4 (https://github.com/mauriceling/dose/releases/tag/v1.0.4), which is licensed under GNU General Public License version 3, except for COPADS (Collection of Python Algorithms and Data Structures, which is released under Python Software Foundation License version 2). DOSE codebase is hosted and available for forking at https://github.com/mauriceling/dose.

The rest of the manuscript is organized as follows. Section 2 provides a brief code description of DOSE version 1.0.4 (for detailed description, please see (Castillo and Ling, 2014a; Lim et al., 2010; Ling, 2012b)). Section 3 presents all 5 new and improved codes of DOSE; namely, *dose.py*, *simulation_calls.py*, *database_calls.py*, *dose_world.py* (bug fixed; original version in (Ling, 2012a)) and *genetic.py* (original version in (Lim et al., 2010)). Two other code files (original version in (Ling, 2012b)); namely, *register_machine.py* and *ragaraja.py*; were unchanged from the original version and will not be presented. Section 4 presents the simulation implementations for the first case study described (Castillo and Ling, 2014a), which examined the effects of different migration schemes on genetic heterozygosity (local genetic distance). Section 5 presents an example of reviving a simulation from Section 4 from the database file and continuing the simulation for another 200 generations. A brief guide on how to write a simulation is provided to end this article.

CODE DESCRIPTION

DOSE version 1.0.4 builds on on previous versions of DOSE (Castillo and Ling, 2014a; Ling, 2012a; Ling, 2012b), which used a previous implementation of GA (*genetic.py*; (Lim et al., 2010)). The architecture of DOSE version 1.0.4 is given in Figure 1.

The most important enhancement to the original implementation of DOSE (Ling, 2012a) is the ability for the simulations to log all events into a SQLite database and subsequently, the ability to revive a simulation from the logged records and continuing the simulation. This is achieved in *database_calls.py* by logging all data attributes of each organism and the entire ecological system into the SQLite database, as well as re-creating all simulation objects (organisms and ecosystem) at a specific generation from the database. Other enhancements (in GA library; *genetic.py*; original version in Lim et al., 2010) are as follows: firstly, each organism is given a 32-character randomly generated name at initialization to aid in ancestry tracking. Secondly, chromosomes will inherit parental background mutation after crossover. Lastly, the statuses of each organism are increased from 6 (Lim et al., 2010) to 12. New statuses are

- parents (identity of parents),
- gender (gender of organism),
- blood (result of genomic interpretation or expression by Rajaraga interpreter),
- identity (32-character randomly generated name),
- deme (defined as a sub-population or local population), and
- location (location of the organism within the ecosystem).

The main application programming interface (API) is provided by *dose.py*, which contains the main functions needed to write a simulation. Once all needed simulation parameters and functions are defined (for details, please see Castillo and Ling, 2014a), *dose.simulate* function is called (see examples in Section 4). This will in turn call functions in *simulation_calls.py* prepare the simulation (Figure 2) by constructing, initializing the DOs, and mapping each DO onto the world. Simulation preparation is dependent on whether the simulation is a new simulation (regular simulation) or revival and continuation of an existing simulation (simulation revival) from database or files. For regular simulation, if the initial chromosome/genome is given, it will be used for organism contruction; otherwise, a chromosome of single base will be used. The constructed organisms (population(s)) will be deployed accordingly before running the simulation (simulation core in Figure 2). On the other hand, a simulation can only be revived if either the simulation log files or the simulation results database with all required parameters and attributes logged are present. These will be used to reconstruct the state of each organism, the world, and mapping each organism to the ecosystem before running the simulation.

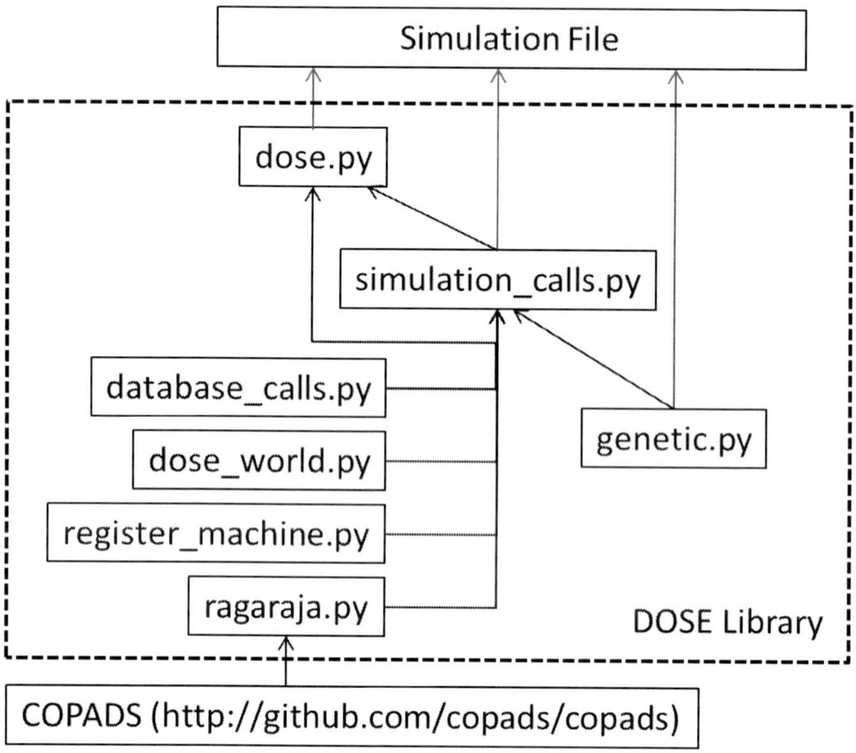

Figure 1. Architecture of DOSE Library. DOSE consists of 7 main files. Four of the files; *database_calls.py*, *dose_world.py*, *register_machine.py* and *ragaraja.py*; were used by *dose.py* and *simulation_calls.py*. *genetic.py* provides the GA for simulation. COPADS (released under Python Software Foundation License version 2) is required by Rajaraja interpreter (*ragaraja.py*; Ling, 2012b). Simulation File represents the simulation implementation, such as those in Sections 4 and 5 and requires *dose.py* and *simulation_calls.py*. *genetic.py*.

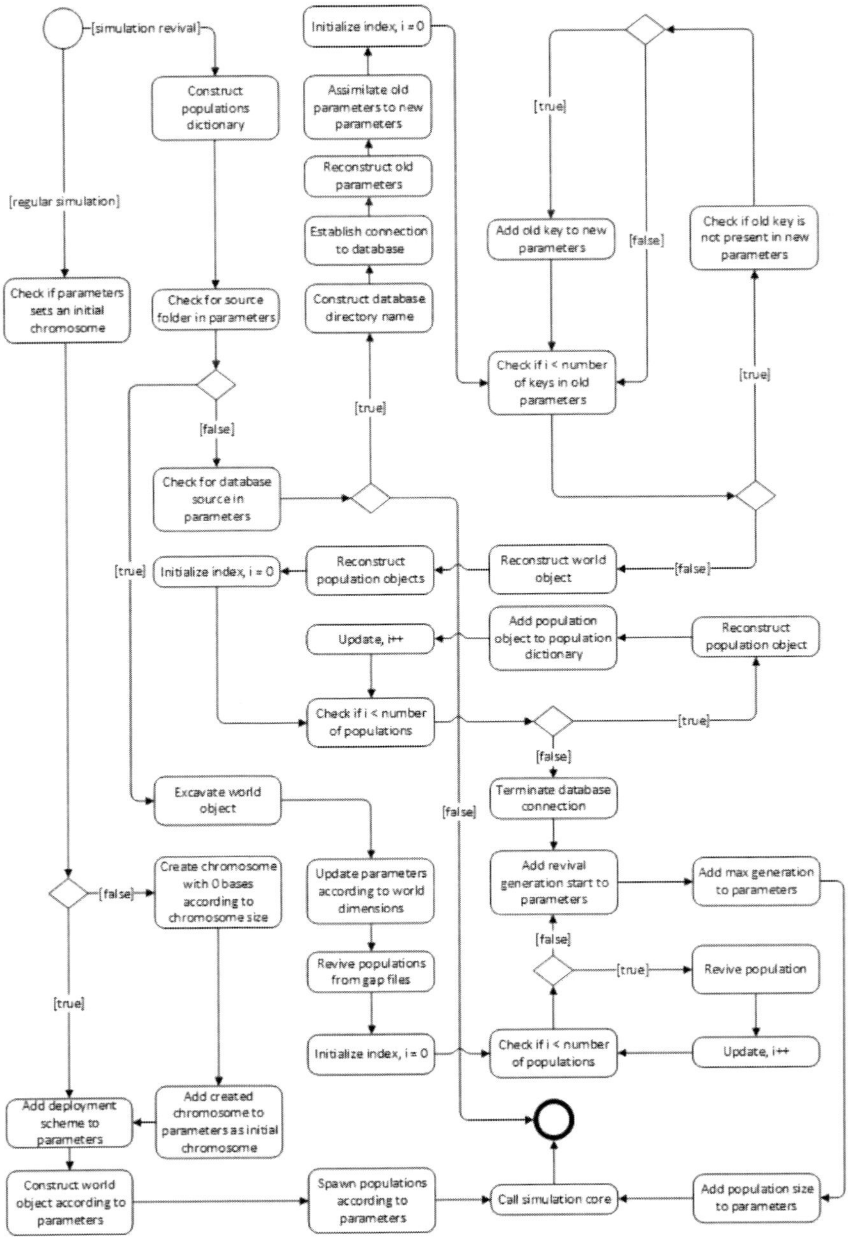

Figure 2. Activity Diagram for Simulation Preparation. Simulation preparation procedure is dependent on whether the simulation is a new simulation (regular simulation) or revival and continuation of an existing simulation (simulation revival) from database or files.

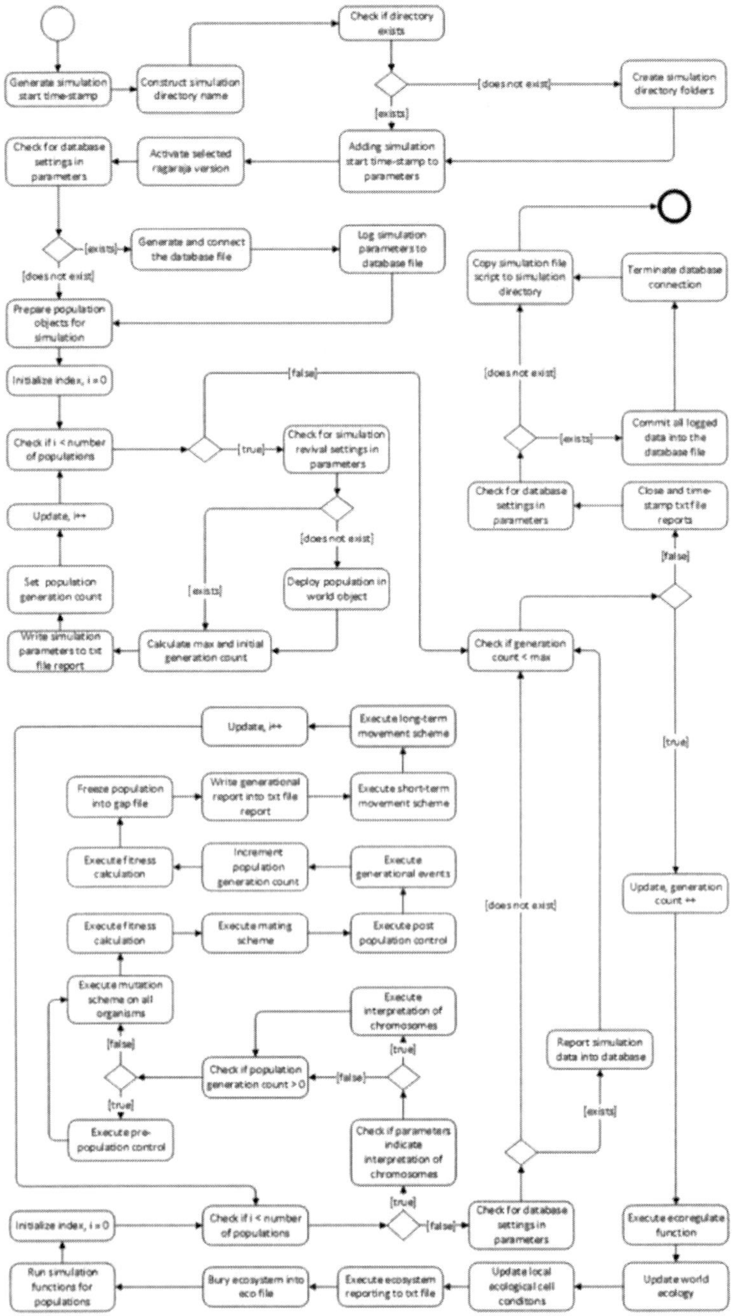

Figure 3. Activity Diagram for Simulation Execution.

After simulation preparation is completed, a unique folder for storing text-based simulation results (results folder) is created using the date/time stamp of the simulation execution. Logging database will be created if database logging is requested and the database file is not present. The simulation is then executed from first generation to the maximum generation as defined in the parameter, and the events are reported into a text file or database as required (Figure 3; Castillo and Ling, 2014a). After DO initialization, the current simulation driver simulates each organism and ecological cell sequentially (Ling, 2012a). Hence, *simulation_calls.py* performs the task of the original command line simulation runner, *run_dose.py* (Ling, 2012a).

FORMATTED DOSE CODE FILES[1]

File Name: metadata.py

```
'''Digital Organism Simulation Environment (DOSE) Metadata

Date created: 14th February 2014'''

pypi_name = 'digital organism simulation environment'
description = '''Digital Organism Simulation \
Environment (DOSE)'''
version = '1.0.4'
maintainer = 'Maurice HT Ling'
email = 'mauriceling@acm.org'
download_url = \
'https://github.com/mauriceling/dose/releases'
project_start_year = '2010'
project_website = 'https://github.com/mauriceling/dose'
documentation_website = \
'http://maurice.vodien.com/project-dose'
short_license = 'GNU General Public License version 3'
long_license = '''Unless otherwise specified, dose.copads
package will be licensed under Python Software Foundation
```

[1] These code files had been formatted for display and printing purposes; hence, different from the original codes in http://github.com/mauriceling/dose on a line-by-line basis. However, they are logically, syntactically, and semantically identical to the original codes.

```
License version 2; all other files will be licensed GNU
General Public License version 3.'''
platforms = ['Operating System :: OS Independent']
trove_classifiers = \
    ['Development Status :: 3 - Alpha',
    'Intended Audience :: Developers',
    'Intended Audience :: Education',
    'Intended Audience :: Science/Research',
    'License :: OSI Approved :: GNU General Public \
    License v3 (GPLv3)',
    'Natural Language :: English',
    'Operating System :: OS Independent',
    'Programming Language :: Python :: 2 :: Only',
    'Topic :: Education',
    'Topic :: Scientific/Engineering :: Artificial Life',
    'Topic :: Scientific/Engineering :: Bio-Informatics',
    'Topic :: Software Development :: Libraries :: \
    Python Modules']

from datetime import datetime
copyright_year = '-'.join([project_start_year,
                           str(datetime.now().year)])

long_description = '''%s

Life is fascinating and deeply intriguing. Despite so,
life forms on Earth or carbon-based life forms as a group
is just one form, one possible sample of possibly a whole
magnitude of life. Even then, there are many aspects of
life that cannot be deciphered even by examining current
life forms; for example, how did chemical reactions
organize themselves into biochemical pathways? How did
life start? How is intelligence formed?

To answer such questions, we will have to restart our
evolutionary time to the very beginning - clearly an
impossibly gargantuan task. At the same time, studying
biological/carbon-based life forms is expensive, time
consuming and destructive. As a molecular biologist, there
is no way I can examine the entire genome of even a
bacteria in an inanimate state, then somehow allow it to
continue living as if time had just stopped while I am
examining it.

However, if I can simulate a bacteria or any life form in
a computer, then I can make a digital copy of the
bacterium, pull it apart to study it while the original
bacterium continues "living" in my virtual world without
```

even knowing that it had been duplicated. Many biologists thought of virtual life forms as a new way to learn about life itself. Studying of virtual life forms is known as Artificial Life and I term "virtual life forms" as "digital organisms". There are several advantages in experimenting using digital organisms. Firstly, generation time can be much faster compared to most biological life. Secondly, it is usually cheaper to examine computer simulations than working on actual biological life. Perhaps the most important advantage of looking at life from this perspective is that by recreating life in a different medium, we are not limited to our own system of carbon-based life; hence, studying life as what-it-could-be.

Digital Organisms Simulation Environment (DOSE) is essentially a virtual world simulator for studying digital organisms. I will argue that digital organisms are considered living organisms (Koh and Ling, 2013). Despite so, being a molecular biologist by training, I have a hard time mapping components of digital organisms into biological life whenever such components are too abstract.

Hence, I decided to design an artificial life / digital organism simulator that bears resemblance to biological life and ecology. These are the foundation papers:

1. Lim, JZR, Aw, ZQ, Goh, DJW, How, JA, Low, SXZ, Loo, BZL, Ling, MHT. 2010. A genetic algorithm framework grounded in biology. The Python Papers Source Codes 2: 6.

This manuscript describes the implementation of a GA framework that uses biological hierarchy - from chromosomes to organisms to population.

2. Ling, MHT. 2012. An Artificial Life Simulation Library Based on Genetic Algorithm, 3-Character Genetic Code and Biological Hierarchy. The Python Papers 7: 5.

Genetic algorithm (GA) is inspired by biological evolution of genetic organisms by optimizing the genotypic combinations encoded within each individual with the help of evolutionary operators, suggesting that GA may be a suitable model for studying real-life evolutionary processes. This paper describes the design of a Python library for artificial life simulation, Digital Organism Simulation Environment (DOSE), based on GA and biological hierarchy starting from genetic sequence to population. A

```
3-character instruction set that does not take any operand
is introduced as genetic code for digital organism. This
mimics the 3-nucleotide codon structure in naturally
occurring DNA. In addition,
the context of a 3-dimensional world composing of
ecological cells is introduced to simulate a physical
ecosystem.

3. Ling, MHT. 2012. Ragaraja 1.0: The Genome Interpreter
of Digital Organism Simulation Environment (DOSE). The
Python Papers Source Codes 4: 2.

This manuscript describes the implementation and test of
Ragaraja instruction set version 1.0, which is the core
genomic interpreter of DOSE.

From this foundation, the complete suite of Digital
Organisms Simulation Environment (DOSE) can be build.

Project website: U{%s}

Documentation can be found at %s

License: %s

Copyright %s, Maurice HT Ling (on behalf of all authors).
''' % (description,
       project_website,
       documentation_website,
       long_license,
       copyright_year)
```

File Name: __init__.py

```
'''Digital Organism Simulation Environment (DOSE)

Date created: 27th September 2013'''
import metadata

__version__ = metadata.version
__maintainer__ = metadata.maintainer
__email__ = metadata.email
__description__ = metadata.long_description

# Package imports (in ascending order of package names)
```

```
import copads

# Module imports (in ascending order of module names)
import database_calls
import dose
import genetic
import register_machine
import ragaraja

# COPADS Class imports (in ascending order of module
# names, then class names)
from copads.lindenmayer import lindenmayer

# DOSE Class imports (in ascending order of module names,
# then class names)
from dose import dose_functions
from dose_world import World
from genetic import Chromosome
from genetic import Organism
from genetic import Population

# Function imports (in ascending order of module names,
# then function names)
from database_calls import connect_database
from database_calls import db_list_datafields
from database_calls import db_list_generations
from database_calls import db_list_simulations
from database_calls import db_list_population_name
from database_calls import db_get_ecosystem
from database_calls import \
db_get_organisms_chromosome_sequences
from database_calls import db_get_organisms_genome
from database_calls import db_get_organisms_status
from dose import database_report_populations
from dose import database_report_world
from dose import filter_age
from dose import filter_deme
from dose import filter_gender
from dose import filter_location
from dose import filter_status
from dose import filter_vitality
from dose import revive_simulation
from dose import simulate
from genetic import crossover
from genetic import population_constructor
from genetic import population_simulate
```

File Name: dose.py

```
'''Application Programming Interface (API) for DOSE
(digital organism simulation environment). This contains
the main functions and operations needed to write a DOSE
simulation. This file will be imported as top level (from
dose import *) when DOSE is imported; hence, all functions
in this file can be assessed at top level.
Date created: 27th September 2013'''
import sys, os, random, inspect

import ragaraja, register_machine
import dose_world
import genetic, simulation_calls

from simulation_calls import spawn_populations
from simulation_calls import simulation_core
from simulation_calls import excavate_world
from simulation_calls import revive_population

from database_calls import connect_database
from database_calls import \
db_reconstruct_simulation_parameters
from database_calls import db_reconstruct_population
from database_calls import db_reconstruct_world
from database_calls import db_list_simulations

class dose_functions():
'''Abstract class to contain all of the simulation
specific functions (functions that vary with each
simulation) that are to be defined / implemented by the
user to be used in a simulation. This class should be
inherited by every simulation to over-ride each function /
method.

This set of functions / methods is consolidated functions
/ methods to be over-ridden from genetic.Organism,
genetic.Population, and dose_world.World classes. As a
result, the functions / methods can be working at
different levels - at the level of individual organisms,
at the level of entire population(s), or at the level of
the world.

Please see the examples in examples directory on its use.
'''

    def mutation_scheme(self, organism):
    '''Method / function to trigger mutational events in
```

```
    each chromosome of the genome within an organism.
    This function works at the level of individual
    organisms.

    @param organism: genetic.Organism object
    @return: None'''
        raise NotImplementedError
    def prepopulation_control(self, Populations,
                              pop_name):
    '''Method / function to trigger population control
    events before mating event in each generation. For
    example, it can be used to simulate pre-puberty
    (childhood) death. This function works at the level
    of entire population(s).

    @param Populations: A dictionary containing one or
    more populations where the value is a
    genetic.Population object.
    @param pop_name: Name of the population which is used
    as key in the the dictionary (Populations parameter).
    @return: None'''
        raise NotImplementedError
    def fitness(self, Populations, pop_name):
    '''Method / function to calculate the fitness score
    of each organism within the population(s). This
    function works at the level of entire population(s)
    even though fitness calculation occurs at the
    organism level. The fitness of each organism may be
    stored in Organism.status['fitness'] and may be used
    by mating scheme.

    @param Populations: A dictionary containing one or
    more populations where the value is a
    genetic.Population object.
    @param pop_name: Name of the population which is used
    as key in the the dictionary (Populations parameter).
    @return: None'''
        raise NotImplementedError
    def mating(self, Populations, pop_name):
    '''Method / function to trigger mating events in each
    generation. For example, it can be used to simulates
    mate choices and progeny size. A support function
    provided is genetic.crossover() function which
    generates one random crossover operation between 2
    chromosomes, to simulate meiosis crossover. This
    function may also use one or more of the
    dose.filter_XXX() functions to select or choose
    suitable mates. This function works at the level of
```

```
entire population(s), which means that this function
will have
    - to manage mating scheme and progeny (offspring)
    generation for the entire population
    - add or replace offsprings into the respective
    population(s)
    - (optional) store identity of parent(s) as list;
    for example, [parentA identity, parentB
    identity], in offspring's status['parents'] for
    ancestral tracing.

@param Populations: A dictionary containing one or
more populations where the value is a
genetic.Population object.
@param pop_name: Name of the population which is used
as key in the the dictionary (Populations parameter).
@return: None'''
    raise NotImplementedError
def postpopulation_control(self, Populations,
                           pop_name):
'''Method / function to trigger population control
events after mating event in each generation. For
example, it can be used tosimulate old-age death.
This function works at the level of entire
population(s).

@param Populations: A dictionary containing one or
more populations where the value is a
genetic.Population object.
@param pop_name: Name of the population which is used
as key in the the dictionary (Populations parameter).
@return: None'''
    raise NotImplementedError
def generation_events(self, Populations, pop_name):
'''Method / function to trigger other defined events
in each generation. For example, it can be used to
simulate catastrophe or epidemic that does not occur
regularly, or simulates unusual occurrences of
multiple mutation events. This function works at the
level of entire population(s).

@param Populations: A dictionary containing one or
more populations where the value is a
genetic.Population object.
@param pop_name: Name of the population which is used
as key in the the dictionary (Populations parameter).
@return: None'''
```

```
    raise NotImplementedError
def population_report(self, Populations, pop_name):
'''Method / function to generate a text report of the
population(s) and/or each organisms within the
population at regular intervals, within the
simulation, as determined by "print_frequency" in the
simulation parameters. This function works at the
level of entire population(s).

@param Populations: A dictionary containing one or
more populations where the value is a
genetic.Population object.
@param pop_name: Name of the population which is used
as key in the the dictionary (Populations parameter).
@return: Entire report of a population at a
generation count in a string. To make it human-
readable, usually the report string is both tab-
delimited (for one organism) and newline-delimited
(for entire population).'''
    raise NotImplementedError
def organism_movement(self, Populations, pop_name,
                      World):
'''organism_movement and organism_location are both
methods / functions to execute movement of organisms
within the world. The semantic difference between
organism_movement and organism_location is that
organism_movement is generally used for short travels
while organism_location is used for long travel. For
example, organism_movement can be used to simulate
foraging or nomadic behaviour. This function works at
both the level of entire population(s) and world.

For each organism to move, this function will have to
    - update the number of organisms in each
    World.ecosystem[x-axis][y-axis][z-
    axis]['organisms']
    - update the respective Organism's location in
    the status dictionary (Population[pop_name].
    agents[<index>].status['location'])

@param Populations: A dictionary containing one or
more populations where the value is a
genetic.Population object.
@param pop_name: Name of the population which is used
as key in the the dictionary (Populations parameter).
@param World: dose_world.World object.
@return: None'''
    raise NotImplementedError
```

```
def organism_location(self, Populations,
                      pop_name, World):
'''organism_movement and organism_location are both
methods / functions to execute movement of organisms
within the world. The semantic difference between
organism_movement and organism_location is that
organism_movement is generally used for short travels
while organism_location is used for long travel. For
example, organism_movement can be used to simulate
foraging or nomadic behaviour. This function works at
both the level of entire population(s) and world.

For each organism to move, this function will have to
    - update the number of organisms in each
    World.ecosystem[x-axis][y-axis][z-
    axis]['organisms']
    - update the respective Organism's location in
    the status dictionary (Population[pop_name].
    agents[<index>].status['location'])

@param Populations: A dictionary containing one or
more populations where the value is a
genetic.Population object.
@param pop_name: Name of the population which is used
as key in the the dictionary (Populations parameter).
@param World: dose_world.World object.
@return: None'''
    raise NotImplementedError
def ecoregulate(self, World):
'''Method / function for broad spectrum management of
the entire ecosystem defined as World.ecosystem[x-
axis][y-axis][z-axis]['local_input'] and
World.ecosystem[x-axis][y-axis][z-axis]
['local_output']). For example, it can be used to
simulate temperature, solar radiation, or resource
gradients. This function works at the level of the
world.

@param World: dose_world.World object.
@return: None'''
    raise NotImplementedError
def update_ecology(self, World, x, y, z):
'''Method / function to process the input and output
from the activities of the organisms in the current
ecological cell (defined as World.ecosystem[x-
axis][y-axis][z-axis]['temporary_input'] and
World.ecosystem[x-axis][y-axis][z-
axis]['temporary_output']) into a local ecological
```

```
cell condition (defined as World.ecosystem
[x-axis][y-axis][z-axis]['local_input'] and
World.ecosystem[x-axis][y-axis][z-
axis]['local_output']), and update the ecosystem
(which is essentially the ecological cell adjacent to
World.ecosystem[x-axis][y-axis][z-axis]). For
example, it can be used to simulate secretion of
chemicals or use of resources (such as food) by
organisms, and diffusion of secretions to the
neighbouring ecological cells.

Essentially, this function simulates the "diffusion"
of local situation outwards. This function works at
the level of the world.

@param World: dose_world.World object.
@param x: x-axis of the World.ecosystem to identify
the cell.
@param y: y-axis of the World.ecosystem to identify
the cell.
@param z: z-axis of the World.ecosystem to identify
the cell.
@return: None'''
    raise NotImplementedError
def update_local(self, World, x, y, z):
'''Method / function to update local ecological cell
condition (defined as World.ecosystem[x-axis][y-
axis][z-axis]['local_input'] and World.ecosystem[x-
axis][y-axis][z-axis]['local_output']) from the
ecosystem (World.ecosystem [x-axis][y-axis][z-
axis]['local_input'] and World.ecosystem[x-axis][y-
axis][z-axis]['local_output'] of adjacent ecological
cells). For example, it can be used to simulates
movement or diffusion of resources from the ecosystem
to local.

Essentially, this function is the reverse of
update_ecology() function. In this case, the local
ecological cell is affected by adjacent conditions.
This function works at the level of the world.

@param World: dose_world.World object.
@param x: x-axis of the World.ecosystem to identify
the cell.
@param y: y-axis of the World.ecosystem to identify
the cell.
@param z: z-axis of the World.ecosystem to identify
the cell.
```

```
@return: None'''
    raise NotImplementedError
def report(self, World):
'''Method / function to generate a text report of the
ecosystem status (World.ecosystem) at regular
intervals, within the simulation, as determined by
"print_frequency" in the simulation parameters. This
function works at the level of world.

@param World: dose_world.World object.
@return: Entire report of a population at a
generation count in a string. To make it human-
readable, usually the report string is both tab-
delimited (for one organism) and newline-delimited
(for entire population).'''
    raise NotImplementedError
def deployment_scheme(self, Populations, pop_name,
                      World):
'''Method / function to implement a user-specific /
simulation-specific deployment scheme used to deploy
organisms into the World. This function will only be
used when "deployment_code" in simulation parameters
dictionary equals to 0. This function works at all
three levels - organism(s), population(s), and world.

@param Populations: A dictionary containing one or
more populations where the value is a
genetic.Population object.
@param pop_name: Name of the population which is used
as key in the the dictionary (Populations parameter).
@param World: dose_world.World object.
@return: None'''
    raise NotImplementedError
def database_report(self, con, cur, start_time,
                    Population, World,
                    generation_count):
'''Method / function to implement database logging of
each organism in each population, and the ecosystem
status. The frequency of logging is determined by
"database_logging_frequency" in the
simulation_parameters.

Three database tables had been defined for use in
this function:
    - organisms where the structure is (start_time
    text, pop_name text, org_name text, generation
    text, key text, value text)
    - world where the structure is (start_time text,
```

```
    x text, y text, z text, generation text, key
    text, value text)
    - miscellaneous where the structure is
    (start_time text, generation text, key text,
    value text)

The logical purpose of these tables are:
    - organisms, to log status of each organism.
    Starting time of current simulation (start_time),
    population name (pop_name), organism name
    (org_name), and generation count (generation)
    are used as complex primary key to identify the
    organism within a specific simulation at a
    specific generation. Key and value pair makes up
    the actual data to be logged where key is the
    field and value is the datum or attribute.
    - world, to log the status of the ecosystem.
    Starting time of current simulation (start_time),
    location of ecological cell (x, y, z as
    coordinates), and generation count (generation)
    are used as complex primary key to identify the
    ecological cell within a specific simulation at a
    specific generation. Key and value pair makes up
    the actual data to be logged where key is the
    field and value is the datum or attribute.
    - miscellaneous, is used to log other undefined
    data. Starting time of current simulation
    (start_time), and generation count (generation)
    are used as complex primary key to identify a
    specific generation within a specific simulation.
    Key and value pair makes up the actual data to be
    logged where key is the field and value is the
    datum or attribute.

@param con: Database connector. See Python DB-API for
details.
@param cur: Database cursor. See Python DB-API for
details.
@param start_time: Starting time of current
simulation in the format of <date>-<seconds since
epoch>; for example, 2013-10-11-1381480985.77.
@param Population: A dictionary containing one or
more populations where the value is a
genetic.Population object.
@param World: dose_world.World object.
@param generation_count: Current number of
generations simulated.
@return: None'''
```

```
        raise NotImplementedError
def database_report_populations(con, cur, start_time,
                                Populations,
                                generation_count):
'''Function to log organisms' status and genome into
database. Organisms' status is implemented as a dictionary
and each key-value pair in the status is logged as a
separate record. Similarly, each chromosome is logged as a
separately record. A combination of starting time of the
simulation, population name, organism's name, and the
generation can identify all the data specific to an
organism within a population, at a specific generation
within a simulation. This function is a complete logger -
it logs everything there is about an organism. There is
nothing to log for populations.

The following transformations of data are made:
    - Organism.status['blood'] is a list of numbers
    resulting from interpreting the genome by Ragaraja
    interpreter. The numbers are concatenated and
    delimited by '|'. For example, [1, 2, 3] ==> 1|2|3
    - Organism.status['location'] is a tuple of 3
    integers (x, y, z) forlocation of ecological cell.
    The numbers are concatenated and delimited by '|'.
    For example, (2, 3, 4) ==> 2|3|4
    - Each chromosome in Organism.genome is a list of
    bases. These bases are concatenated with no
    delimiter. For example, [1, 2, 3] ==> 123

@param con: Database connector. See Python DB-API for
details.
@param cur: Database cursor. See Python DB-API for
details.
@param start_time: Starting time of current simulation in
the format of <date>-<seconds since epoch>; for example,
2013-10-11-1381480985.77.
@param Populations: A dictionary containing one or more
populations where the value is a genetic.Population
object.
@param generation_count: Current number of generations
simulated.
@return: None'''
    generation = str(generation_count)
    for pop_name in list(Populations.keys()):
        for org in Populations[pop_name].agents:
            org_name = str(org.status['identity'])
            # log each item in Organism.status dictionary
            for key in [key for key in
```

```
                list(org.status.keys())
                if key != 'identity']:
                if key in ('blood', 'location',
                           'parents'):
                    try:
                        # TypeError will occur when
                        # genome/chromosomes are
                        # not interpreted
                        value = '|'.join([str(x)
                               for x inorg.status[key]])
                    except TypeError: value = ''
                else:
                    value = str(org.status[key])
                cur.execute('insert into organisms values
                    (?,?,?,?,?,?)',
                    (str(start_time), str(pop_name),
                     org_name, generation, key, value))
            # log each chromosome sequence
            for chromosome_count in
                range(len(org.genome)):
                key = 'chromosome_' + \
                      str(chromosome_count)
                sequence = \
            ''.join(org.genome[chromosome_count].sequence)
                cur.execute('insert into organisms values
                    (?,?,?,?,?,?)',
                    (str(start_time), str(pop_name),
                     org_name, generation, key,
                     sequence))
    con.commit()

def database_report_world(con, cur, start_time, World,
                          generation_count):
'''Function to log World.ecosystem into database. The
ecosystem is made up of a collection of ecological cells,
identified by the (x, y, z) coordinates within the
ecosystem. Each ecological cell is implemented as a
dictionary of ecological status. This function logs the
entire set of ecological cells; thus, a complete logger.
Each ecosystem within a specific simulation, at a specific
time/generation, can be identified by a combination of
starting time of simulation and generation. Each
ecological cell within the ecosystem can be further
identified by the (x, y, z) coordinates.

@param con: Database connector. See Python DB-API for
details.
```

```
@param cur: Database cursor. See Python DB-API for
details.
@param start_time: Starting time of current simulation in
the format of <date>-<seconds since epoch>; for example,
2013-10-11-1381480985.77.
@param World: dose_world.World object.
@param generation_count: Current number of generations
simulated.
@return: None'''
    generation = str(generation_count)
    ecosystem = World.ecosystem
    location = [(x, y, z)
                for x in range(len(ecosystem))
                    for y in range(len(ecosystem[x]))
                        for z in
                            range(len(ecosystem[x][y]))]
    for cell in location:
        eco_cell = ecosystem[cell[0]][cell[1]][cell[2]]
        for key in list(eco_cell.keys()):
            value = str(eco_cell[key])
            cur.execute('insert into world values
                        (?,?,?,?,?,?,?)',
                        (str(start_time), str(cell[0]),
                         str(cell[1]), str(cell[2]),
                         generation, key, value))
    con.commit()

def filter_deme(deme_name, agents):
'''Function to identify organisms (agents) with a specific
sub-population name (also known as deme) within a
population. Demes can be considered as strains in
bacteria, breed or sub-species in animals, races in
humans, and cultivars in plants. This function is can be
used to support the identification of suitable mates for
mating schemes.

@param deme_name: Name of deme (sub-population name)
@type deme_name: string
@param agents: A list of organisms, such as
Population.agents.
@return: List of Organism objects'''
    extract = [individual for individual in agents
               if individual.status['deme'].upper() == \
               deme_name.upper()]
    return extract

def filter_gender(gender, agents):
```

```
'''Function to identify organisms (agents) with a specific
gender within a population. This function is can be used
to support the identification of suitable mates for mating
schemes.

@param gender: Gender
@type gender: string
@param agents: A list of organisms, such as
Population.agents.
@return: List of Organism objects'''
    extract = [individual for individual in agents
               if individual.status['gender'].upper() ==\
               gender.upper()]
    return extract

def filter_age(minimum, maximum, agents):
'''Function to identify organisms (agents) within a
certain age range in a population. This function is can be
used to support the identification of suitable mates for
mating schemes.

@param minimum: Minimum age
@type minimum: float
@param maximum: Maximum age
@type maximum: float
@param agents: A list of organisms, such as
Population.agents.
@return: List of Organism objects'''
    extract = [individual for individual in agents
               if float(individual.status['age']) > \
               (float(minimum) - 0.01) \
               and float(individual.status['age']) < \
               float(maximum) + 0.01]
    return extract

def filter_location(location, agents):
'''Function to identify organisms (agents) of a population
within a specific ecological cell. This function is can be
used to support the identification of suitable mates for
mating schemes.

@param location: (x, y, z) coordinates within the
World.ecosystem
@param location: tuple
@param agents: A list of organisms, such as
Population.agents.
@return: List of Organism objects'''
    extract = [individual for individual in agents
```

```
                if individual.status['location'] == \
                location]
    return extract

def filter_vitality(minimum, maximum, agents):
'''Function to identify organisms (agents) within a
certain vitality score in a population. This function is
can be used to support the identification of suitable
mates for mating schemes.

@param minimum: Minimum vitality score
@type minimum: float
@param maximum: Maximum vitality score
@type maximum: float
@param agents: A list of organisms, such as
Population.agents.
@return: List of Organism objects'''
    extract = [individual for individual in agents
            if float(individual.status['vitality']) > \
            (float(minimum) - 0.01) \
            and float(individual.status['vitality']) < \
            float(maximum) + 0.01]
    return extract

def filter_status(status_key, condition, agents):
'''Generic function to identity organisms (agents) within
a population via the status of organisms (Organism.status
dictionary). This function is can be used to support the
identification of suitable mates for mating schemes.

@param status_key: Status (key in Organism.status
dictionary) to filter
@type status_key: string
@param condition: Condition of the status (value in
Organism.status dictionary) to filter. This can be a
unique condition, such as "True", or a range, such as
(minimum, maximum) to define the minimum and maximum value
of the condition.
@param agents: A list of organisms, such as
Population.agents.
@return: List of Organism objects'''
    if type(condition) in (str, int, float, bool):
        extract = [individual for individual in agents
                   if individual.status[status_key] == \
                   condition]
    else:
        extract = [individual for individual in agents
        if float(individual.status[status_key]) > \
```

```
          float(condition[0]) - 0.01 \
        and float(individual.status[status_key]) < \
          float(condition[1]) + 0.01]
    return extract

def revive_simulation(rev_parameters, sim_functions):
    print('\n[' + \
    rev_parameters["simulation_name"].upper() + \
    ' REVIVAL SIMULATION]')
    Populations = {}
    if "sim_folder" in rev_parameters:
        print('Accessing simulation files directory...')
        print('Excavating World entity: ' + \
              rev_parameters['eco_file'] + '...')
        World = \
        excavate_world(rev_parameters['sim_folder'] + \

        rev_parameters['eco_file'])
        print('Updating parameters with World \
        dimensions...')
        rev_parameters["world_z"] = \
            len(World.ecosystem[0][0][0])
        rev_parameters["world_y"] = \
            len(World.ecosystem[0][0])
        rev_parameters["world_x"] = \
            len(World.ecosystem[0])
        for i in range(len(rev_parameters["pop_files"])):
            print('\nReviving population file: ' + \
                rev_parameters["pop_files"][i] + '...')
            pop_file = rev_parameters["sim_folder"] + \
                rev_parameters["pop_files"][i]
       Populations[rev_parameters["population_names"][i]]
        = revive_population(pop_file)
        print('\nUpdating revival generation start in \
        simulation parameters...')
        rev_parameters["rev_start"] = \
            [Populations[pop_name].generation
              for pop_name in Populations]
    elif "database_source" in rev_parameters:
        print('Constructing database directory...')
        dbpath = os.sep.join([os.getcwd(),
                 'Simulations',
                 rev_parameters["database_source"]])
        print('Connecting to database file: ' + \
            rev_parameters["database_source"] + '...')
        (con, cur) = connect_database(dbpath, None)
        if rev_parameters["simulation_time"] == \
            'default':
```

```
            print('Acquiring simulation starting \
                 time...')
            rev_parameters["simulation_time"] = \
                db_list_simulations(cur)[0][0]
        print('Reconstructing old simulation \
              parameters...')
        temp_parameters = \
        db_reconstruct_simulation_parameters(cur,
            rev_parameters["simulation_time"])
        print('Assimilating old simulation parameters \
              with new simulation parameters...')
        for key in temp_parameters:
            if key not in rev_parameters:
                rev_parameters[key] = \
                    temp_parameters[key]
        print('Reconstructing World entity...')
        World = db_reconstruct_world(cur,
                rev_parameters["simulation_time"],
                rev_parameters["rev_start"][0])
        print('\nUpdating population names parameter...')
        for pop_name in \
           rev_parameters["population_names"]:
            print('Reconstructing population: ' + \
                 pop_name + '...')
            Populations[pop_name] = \
            db_reconstruct_population(cur,
            rev_parameters["simulation_time"], pop_name,
            rev_parameters["rev_start"][rev_parameters
            ["population_names"].index(pop_name)])
        print('Terminating database connection...')
        con.close()
    print('Updating last generation revival and \
    population size simulation parameters...')
    rev_parameters["rev_finish"] = \
    [(Populations[pop_name].generation + \
     rev_parameters["extend_gen"])
     for pop_name in Populations]
    rev_parameters["rev_pop_size"] = \
    [len(Populations[pop_name].agents)
     for pop_name in Populations]
    print('\nStarting simulation core...')
    simulation_core(sim_functions, rev_parameters,
                    Populations, World)

def simulate(sim_parameters, sim_functions):
'''Function called by simulation to run the actual
simulation based on a set of parameters and functions.
```

```
Simulation parameters dictionary contains the following
keys:
    - simulation_name: Short name of the simulation.
    - population_names: Population name(s) in a list.
    - population_locations: A list of lists containing
    (x, y, z) coordinates in the world to deploy the
    populations. There must be equal numbers of items in
    this list as population_names. However, one
    population can be deployed into one or more
    ecological cell(s).
    - deployment_code: Defines the type of deployment.
    Allowable values are 0 (custom deployment scheme
    which users can implement by over-riding
    dose.dose_functions.deployment_scheme() function), 1
    (all organisms are in one location), 2 (organisms are
    randomly deployed across a list of population
    locations given for the population), 3 (oganisms are
    randomly deployed across a list of population
    locations given for the population), 4 (organisms are
    dispersed from a specific location where the defined
    location will have the maximum allocation before
    dispersal happens). (see simulation_calls.deployment
    for more details)
    - chromosome_bases: List containing allowable bases
    in the chromosome.
    - background_mutation: Defines background mutation
    rate where 0.01 represents 1% mutation and 0.5
    represents 50% mutation.
    - additional_mutation: Defines mutation rate on top
    of background mutation rate where 0.01 represents 1%
    mutation and 0.5 represents 50% mutation.
    - mutation_type: Type of mutation for background
    mutation rate. Allowable values are 'point' (point
    mutation), 'insert' (insert a base), 'delete' (delete
    a base), 'invert' (invert a stretch of the
    chromosome), 'duplicate' (duplicate a stretch of the
    chromosome), 'translocate' (translocate a stretch of
    chromosome to another random position).
    - chromosome_size: Number of bases for a chromosome.
    - genome_size: Number of chromosome(s) in a genome.
    - max_tape_length: Maximum number of cells for array
    used in Ragaraja interpreter, which will be stored as
    Organism.status['blood'].
    - clean_cell: Toggle to define if a new tape will be
    used for Ragaraja interpreter. If True, at every
    Ragaraja interpretation of the source (genome), a new
    tape ([0] * max_tape_length) will be provided. If
    False, the tape stored as Organism.status['blood']
```

```
(results from previous interpretation of genome) will
be used.
- interpret_chromosome: Toggle to define whether
genomes will be interpreted by Ragaraja. If True,
chromosomes of each organism will be interpreted by
Ragaraja.
- max_codon: Maximum number of Ragaraja instructions
to execute per organism, which can be used as force-
break from endless loop.
- population_size: Number of organisms per population
for initial deployment.
- eco_cell_capacity: Maximum number of organisms per
ecological cell at the time of deployment prior to
the start of simulation.
- world_x: Number of ecological cells in the x-axis
of the World.ecosystem.
- world_y: Number of ecological cells in the y-axis
of the World.ecosystem.
- world_z: Number of ecological cells in the z-axis
of the World.ecosystem.
- goal: Goal for population to reach. This provides a
goal for use in fitness functions.
- maximum_generations: Number of generations to
simulate.
- fossilized_ratio: Proportion (less than 1) of the
population to fossilize or freeze.
- fossilized_frequency: Number of generations
(intervals) for each population fossilization or
freezing event.
- print_frequency: Number of generations (intervals)
for each reporting event into files.
- ragaraja_version: Ragaraga instruction version to
activate. (see ragaraja.activate_version() for more
details)
- ragaraja_instructions: A list defining a set of
Ragaraga instructions to activate. This is only
useful when ragaraja_version is 0.
- eco_buried_frequency: Number of generations
(intervals) for each ecological freezing or burial
event.
- database_file: Logging database file name. This
file will be located in <current working
directory>/Simulations folder
- database_logging_frequency: Number of generations
(intervals) for each database logging event.
```

```
Methods / Functions from dose.dose_functions class to be
over-ridden as simulation_functions (for more details,
please look at dose.dose_functions class):
    - organism_movement: organism_movement and
    organism_location are both methods / functions to
    execute movement of organisms within the world. The
    semantic difference between organism_movement and
    organism_location is that organism_movement is
    generally used for short travels while
    organism_location is used for long travel.
    - organism_location: organism_movement and
    organism_location are both methods / functions to
    execute movement of organisms within the world. The
    semantic difference between organism_movement and
    organism_location is that organism_movement is
    generally used for short travels while
    organism_location is used for long travel.
    - ecoregulate: Broad spectrum management of the
    entire ecosystem.
    - update_ecology: Process the input and output from
    the activities of the organisms in the current
    ecological cell into a local ecological cell
    condition, and update the ecosystem.
    - update_local: Update local ecological cell
    condition from the ecosystem.
    - report: Generate a text report of ecosystem
    (World.ecosystem) at regular intervals, within the
    simulation, as determined by "print_frequency" in the
    simulation parameters.
    - fitness: Calculate the fitness score of each
    organism within the population(s).
    - mutation_scheme: Trigger mutational events in each
    chromosome of the genome within an organism.
    - prepopulation_control: Trigger population control
    events before mating event in each generation.
    - mating: Trigger mating events in each generation.
    - postpopulation_control: Trigger population control
    events after mating event in each generation.
    - generation_events: Trigger other defined events in
    each generation.
    - population_report: Generate a text report of the
    population(s) and/or each organisms within the
    population at regular intervals, within the
    simulation, as determined by "print_frequency" in the
    simulation parameters
    - database_report: Implement database logging of each
    organism in each population, and the ecosystem
    status. The frequency of logging is determined by
```

```
    "database_logging_frequency" in the
    simulation_parameters.
    - deployment_scheme: Implement a user-specific /
    simulation-specific deployment scheme used to deploy
    organisms into the World. This function will only be
    used when "deployment_code" in simulation parameters
    dictionary equals to 0.

@param sim_parameters: Dictionary of simulation parameters
@param sim_functions: A class inherited from
dose.dose_functions class to implement all the needed
simulation functions.'''
    print('\n[' + \
    sim_parameters["simulation_name"].upper() + \
    ' SIMULATION]')
    if "initial_chromosome" not in sim_parameters:
        print('Adding initial chromosome to simulation \
              parameters...')
        sim_parameters["initial_chromosome"] = ['0'] * \
            sim_parameters["chromosome_size"]
    print('Adding deployment scheme to simulation \
    parameters...')
    sim_parameters["deployment_scheme"] = \
        sim_functions.deployment_scheme
    print('Constructing World entity...')
    World = dose_world.World(sim_parameters["world_x"],
                             sim_parameters["world_y"],
                             sim_parameters["world_z"])
    print('Spawning populations...')
    Populations = spawn_populations(sim_parameters)
    print('\nStarting simulation core...')
    simulation_core(sim_functions, sim_parameters,
                    Populations, World)
```

File Name: dose_world.py

```
'''World structure for DOSE (digital organism simulation
environment)
Date created: 13th September 2012

Reference: Ling, MHT. 2012. An Artificial Life Simulation
Library Based on Genetic Algorithm, 3-Character Genetic
Code and Biological Hierarchy. The Python Papers 7: 5.'''
import copy
```

```
class World(object):
'''Representation of a 3-dimensional ecological world.
The ecosystem is made up of ecological cells. Each
ecological cell is modelled as a dictionary of
    - local_input: A list containing processed input,
    representing the partial local ecological condition,
    to be used as input to the organisms in the current
    ecological cell. This is updated by
    World.update_local function.
    - local_output: A list containing processed output,
    representing the partial local ecological condition.
    This is updated by World.update_local function.
    - temporary_input: A list acting as temporary holding
    for input after being fed to the organisms in the
    current ecological cell, which is to be used to
    update local_input and local_output lists by
    World.update_local and World.update_ecology
    functions.
    - temporary_output: A list acting as temporary
    holding for output from the organisms in the current
    ecological cell, which is to be used to update
    local_input and local_output lists by
    World.update_local and World.update_ecology
    functions.
    - organisms: The number of organisms in the current
    ecological cell which is updated by
    World.organism_movement and World.organism_location
    functions.

@see: Ling, MHT. 2012. An Artificial Life Simulation
Library Based on Genetic Algorithm, 3-Character Genetic
Code and Biological Hierarchy. The Python Papers 7: 5.'''

    def __init__(self, world_x, world_y, world_z):
    '''Setting up the world and ecosystem

    @param world_x: number of ecological cells on the x-
    axis
    @type world_x: integer
    @param world_y: number of ecological cells on the y-
    axis
    @type world_y: integer
    @param world_z: number of ecological cells on the z-
    axis
    @type world_z: integer'''
        self.ecosystem = {}

        eco_cell = {'local_input': [],
```

```
                      'local_output': [],
                      'temporary_input': [],
                      'temporary_output': [],
                      'organisms': 0}
        self.world_x = int(world_x)
        self.world_y = int(world_y)
        self.world_z = int(world_z)
        for x in range(self.world_x):
            eco_x = {}
            for y in range(self.world_y):
                eco_y = {}
                for z in range(self.world_z):
                    eco_y[z] = copy.deepcopy(eco_cell)
                eco_x[y] = copy.deepcopy(eco_y)
            self.ecosystem[x] = copy.deepcopy(eco_x)

    def eco_burial(self, filename):
    '''Function to preserve the entire ecosystem.

    @param filename: file name of preserved ecosystem.'''
        try:
            import cPickle as pickle
        except ImportError:
            import pickle

        f = open(filename, 'wb')
        pickle.dump(self.ecosystem, f)
        f.close()

    def eco_excavate(self, filename):
    '''Function to excavate entire ecosystem.

    @param filename: file name of preserved ecosystem.'''
        try:
            import cPickle as pickle
        except ImportError:
            import pickle

        self.ecosystem = \
            pickle.load(open(filename, 'rb'))

    def ecoregulate(self):
    '''Function to simulate events to the entire
    ecosystem. B{This function may be over-ridden by the
    inherited class or substituted to cater for
    ecological schemes but not an absolute requirement
    to do so.}'''
        pass
```

```
def organism_movement(self, x, y, z):
'''Function to trigger organism movement from current
ecological cell to an adjacent ecological cell.
B{This function may be over-ridden by the inherited
class or substituted to cater for mobility schemes
but not an absolute requirement to do so.}

@param x: location of current ecological cell on the
x-axis
@type x: integer
@param y: location of current ecological cell on the
y-axis
@type y: integer
@param z: location of current ecological cell on the
z-axis
@type z: integer'''
    pass
def organism_location(self, x, y, z):
'''Function to trigger organism movement from current
ecological cell to a distant ecological cell. B{This
function may be over-ridden by the inherited class or
substituted to cater for mobility schemes but not an
absolute requirement to do so.}

@param x: location of current ecological cell on the
x-axis
@type x: integer
@param y: location of current ecological cell on the
y-axis
@type y: integer
@param z: location of current ecological cell on the
z-axis
@type z: integer'''
    pass

def update_ecology(self, x, y, z):
'''Function to process temporary_input and
temporary_output from the activities of the organisms
in the current ecological cell into a local
ecological cell condition, and update the ecosystem.
B{This function may be over-ridden by the inherited
class or substituted to cater for ecological schemes
but not an absolute requirement to do so.}

@param x: location of current ecological cell on the
x-axis
@type x: integer
@param y: location of current ecological cell on the
```

```
    y-axis
    @type y: integer
    @param z: location of current ecological cell on the
    z-axis
    @type z: integer'''
        pass

    def update_local(self, x, y, z):
    '''Function to update local ecological cell condition
    from the ecosystem. B{This function may be over-
    ridden by the inherited class or substituted to cater
    for ecological schemes but not an absolute
    requirement to do so.}

    @param x: location of current ecological cell on the
    x-axis
    @type x: integer
    @param y: location of current ecological cell on the
    y-axis
    @type y: integer
    @param z: location of current ecological cell on the
    z-axis
    @type z: integer'''
        pass

    def report(self):
    '''Function to report the status of the world and
    ecosystem. B{This function may be over-ridden by the
    inherited class or substituted to cater for specific
    reporting schemes but not an absolute requirement to
    do so.}

    @return: dictionary of status describing the current
    generation'''
        pass
```

File Name: database_calls.py

```
'''File containing support functions for database logging
of simulations.

Date created: 10th October 2013'''
import os, copy
import sqlite3 as s
```

```
def connect_database(dbpath, sim_parameters=None):
'''Connects to logging database and prepares database for
use, if database does not exist. This function can be used
to connect to the logging database using a file path or
using simulation parameters dictionary.

Simulation parameters dictionary takes precedence - if
simulation parameters dictionary (sim_parameters) is not
None, this function will look for database file in
<simulation execution directory>/Simulations/<database
file>.

In both cases, if the database is not present, it will
create a SQLite3 database file and create the required
database tables.

@param dbpath: File path of logging database. This
parameter will only be used when sim_parameters == None.
@param sim_parameters: Dictionary of simulation
parameters. Default is None.
@return: (con, cur) where
    - con = connector
    - cur = cursor'''
    if sim_parameters:
        dbpath = os.sep.join([os.getcwd(),
                 'Simulations',
                 sim_parameters["database_file"]])
    con = s.connect(dbpath)
    cur = con.cursor()
    cur.execute('''create table if not exists parameters
                (start_time text, simulation_name text,
                key text, value text)''')
    cur.execute('''create table if not exists organisms
                (start_time text, pop_name text,
                org_name text, generation text,
                key text, value text)''')
    cur.execute('''create table if not exists world
                (start_time text, x text, y text,
                z text, generation text,
                key text, value text)''')
    cur.execute('''create table if not exists
                miscellaneous
               (start_time text, generation text,
               key text, value text)''')
    cur.execute('''create index if not exists
                organisms_index1 on organisms
                (pop_name, generation)''')
    cur.execute('''create index if not exists
```

```
                organisms_index2 on organisms
                (generation, org_name)''')
    con.commit()
    return (con, cur)

def db_log_simulation_parameters(con, cur,
                                 sim_parameters):
'''Function to log the simulation parameters. A database
table, parameters, had been defined for simulation
parameters logging where the structure is (start_time
text, simulation_name text, key text, value text). The
starting time of current simulation (start_time) and name
of the current simulation (simulation_name) are used as
complex primary key to identify the current simulation.
All parameters will be logged, except for "starting_time"
and "simulation_name".

The following transformations of data are made:
    - Population name (key = "population_names") is a
    list of names. These are concatenated and delimited
    by '|'. For example, ['pop_01', 'pop_02'] ==>
    pop_01|pop_02
    - Genetic bases (key = "chromosome_bases") is a list
    of bases to make up the genetic code. These are
    concatenated and delimited by '|'. For example, ['1',
    '2'] ==> 1|2
    - Ragaraja instructions to be used (key =
    "ragaraja_instructions") is a list of 3-character
    Ragaraja instructions in numbers. These are
    concatenated and delimited by '|'. For example,
    ['000', '004', '008'] ==> 000|004|008
    - Initial (ancestral) chromosome is a list of bases.
    These bases are concatenated with no delimiter. For
    example, [1, 2, 3] ==> 123

@param con: Database connector from connect_database()
function.
@param cur: Database cursor from connect_database()
function.
@param sim_parameters: Dictionary of parameters used in
simulate on.
@return: (con, cur) where
    - con = connector
    - cur = cursor'''
    start_time = sim_parameters["starting_time"]
    simulation_name = sim_parameters["simulation_name"]
    for key in [k for k in list(sim_parameters.keys())
                if k not in ("simulation_name",
```

```
                                "starting_time")]:
        value = sim_parameters[key]
        if key in ("population_names",
                   "chromosome_bases",
                   "ragaraja_instructions"):
            value = '|'.join([str(x) for x in value])
            cur.execute('''insert into parameters \
                values (?,?,?,?)''', (str(start_time),
                str(simulation_name), str(key), value))
        elif key in ("initial_chromosome"):
            value = ''.join(value)
            cur.execute('''insert into parameters \
                values (?,?,?,?)''', (str(start_time),
                str(simulation_name), str(key), value))
        else:
            cur.execute('''insert into parameters \
                values (?,?,?,?)''', (str(start_time),
                str(simulation_name), str(key),
                str(value)))
    con.commit()
    return (con, cur)

def db_report(con, cur, sim_functions, start_time,
              Populations, World, generation_count):
'''Wrapper around user-implemented database logging which
over-rides dose.dose_functions.database_report() function.

@param con: Database connector from connect_database()
function.
@param cur: Database cursor from connect_database()
function.
@param sim_functions: Object of simulation-specific
functions, which inherits dose.dose_functions class.
@param start_time: Starting time of current simulation in
the format of <date>-<seconds since epoch>; for example,
2013-10-11-1381480985.77.
@param Populations: A dictionary containing one or more
populations where the value is a genetic.Population
object.
@param World: dose_world.World object.
@param generation_count: Current number of generations
simulated.
@return: (con, cur) where
    - con = connector
    - cur = cursor '''
    sim_functions.database_report(con, cur, start_time,
                                  Populations, World,
                                  generation_count)
```

```
    con.commit()
    return (con, cur)

def db_list_simulations(cur, table='parameters'):
'''Function to list simulations, identified by starting
time of the simulation (and simulation name, if
available), from logging database.

@param cur: Database cursor from connect_database()
function.
@param table: Database table name to list simulations.
Allowable values are 'parameters', 'organisms', 'world',
and 'miscellaneous'. Default value is 'parameters'.
@return: A list containing the results.
    - If table = 'parameters', the returned list will be
    a list of list (consisting of [starting time of
    simulation, simulation name]).
    - If table is not 'parameters', the returned list
    will be a list of starting time of simulation.'''
    if table not in ('parameters', 'organisms',
                     'world', 'miscellaneous'):
        table = 'parameters'
    if table == 'parameters':
        cur.execute("""select distinct start_time,
            simulation_name from parameters""")
        return [[str(x[0]), str(x[1])]
                for x in cur.fetchall()]
    else:
        cur.execute("select distinct start_time from %s",
            table)
        return [str(x[0]) for x in cur.fetchall()]

def db_list_generations(cur, start_time,
                        table='organisms'):
'''Function to list all logged generations within a
simulation, identified by starting time of the simulation.

@param cur: Database cursor from connect_database()
function.
@param start_time: Starting time of current simulation in
the format of <date>-<seconds since epoch>; for example,
2013-10-11-1381480985.77.
@param table: Database table name to list simulations.
Allowable values are 'parameters', 'organisms', 'world',
and 'miscellaneous'. Default value is 'organisms'.
@return: A list containing the results (generation
counts).'''
# query plan: SCAN TABLE organisms USING COVERING INDEX
```

```
# organisms_index2
    cur.execute("select distinct generation from %s \
        where start_time='%s'" % (str(table),
        str(start_time)))
    generations = sorted([int(str(x[0]))
                          for x in cur.fetchall()])
    return [str(gen) for gen in generations]

def db_list_datafields(cur, start_time,
                       table='organisms'):
'''Function to list all logged data fields (types of data
logged) within a simulation, identified by starting time
of the simulation.

@param cur: Database cursor from connect_database()
function.
@param start_time: Starting time of current simulation in
the format of <date>-<seconds since epoch>; for example,
2013-10-11-1381480985.77.
@param table: Database table name to list simulations.
Allowable values are 'parameters', 'organisms', 'world',
and 'miscellaneous'. Default value is 'organisms'.
@return: A list containing the results (types of data
logged).'''
    cur.execute("select distinct key from %s where \
        start_time='%s'" % (str(table), str(start_time)))
    return [str(x[0]) for x in cur.fetchall()]

def db_list_population_name(cur, start_time):
'''Function to list all logged populations (as population
names) within a simulation, identified by starting time of
the simulation.

@param cur: Database cursor from connect_database()
function.
@param start_time: Starting time of current simulation in
the format of <date>-<seconds since epoch>; for example,
2013-10-11-1381480985.77.
@return: A list containing the results (types of data
logged).'''
# query plan: SCAN TABLE organisms USING COVERING INDEX
# organisms_index1
    cur.execute("select distinct pop_name from \
        organisms where start_time='%s'" % \
        str(start_time))
    return [str(x[0]) for x in cur.fetchall()]

def db_get_ecosystem(cur, start_time, datafield='all',
```

```
                          generation='all'):
'''Analysis helper function to get a specific field of the
ecosystem (World.ecosystem) or the entire ecosystem for
one or more generations within a simulation (as identified
by the starting time of the simulation).

@param cur: Database cursor from connect_database()
function.
@param start_time: Starting time of current simulation in
the format of <date>-<seconds since epoch>; for example,
2013-10-11-1381480985.77.
@param datafield: The specific datafield in
World.ecosystem to extract. Predefined datafields for
ecosystem are 'local_input', 'local_output',
'temporary_input', 'temporary_output', and 'organisms'.
Default = 'all', which returns the entire
World.ecosystem(s) of the generation(s). Only one or all
datafield(s) in World.ecosystem can be extracted at a
time.
@type datafield: string
@param generation: Generation(s) to extract
World.ecosystem. Default = 'all', which extracts
World.ecosystems for all logged generations within the
specific simulation.
@type generation: list
@return: A dictionary of results - {<generation count>:
<value>}. The value is the value of the specific datafield
if a specific datafield is required (the data type of this
value is dependent on that of the data type of
World.ecosystem[datafield]) or the entire ecosystem as a
dictionary.'''
    if generation == 'all':
        generation = db_list_generations(cur, start_time,
                                         'world')
    else:
        generation = [str(x) for x in generation]
    results = {}
    for gen in generation:
        results[gen] = {}
        ecosystem = db_reconstruct_world(cur, start_time,
                                         gen).ecosystem
        if datafield == 'all':
            results[gen] = ecosystem
        else:
            for x in list(ecosystem.keys()):
                results[gen][x] = {}
                for y in list(ecosystem[x].keys()):
                    results[gen][x][y] = {}
```

```
                    for z in list(ecosystem[x][y].
                    keys()):
                        results[gen][x][y][z] = \
                        ecosystem[x][y][z][datafield]
    return results

def db_get_organisms_status(cur, start_time,
    population_name, datafield='all', generation='all'):
'''Analysis helper function to get a specific field of the
Organism.status dictionary or the entire Organism.status
dictionary for one or more generations within a simulation
(as identified by the starting time of the simulation).

@param cur: Database cursor from connect_database()
function.
@param start_time: Starting time of current simulation in
the format of <date>-<seconds since epoch>; for example,
2013-10-11-1381480985.77.
@param population_name: Name of the population.
@param datafield: The specific key in Organism.status
dictionary to extract. Predefined keys for Organism.status
dictionary are 'age', 'alive', 'blood', 'death', 'deme',
'fitness', 'gender', 'identity',
'lifespan', 'location', 'parents', 'vitality'. Default =
'all', which returns the entire Organism.status
dictionaries of the generation(s). Only one or all
datafield(s) in Organism.status dictionary can be
extracted at a time.
@type datafield: string
@param generation: Generation(s) to extract
Organism.status dictionary. Default = 'all', which
extracts Organism.status dictionary for all logged
generations within the specific simulation.
@type generation: list
@return: A dictionary of dictionary of results -
{<generation count>: {<Organism identity>: <value>}}. The
value is the value of the specific datafield if a specific
datafield is required (the data type of this value is
dependent on that of the data type of
Organism.status[datafield]) or the entire Organism.status
as a dictionary. '''
    if generation == 'all':
        generation = db_list_generations(cur, start_time,
                                         'organisms')
    else:
        generation = [str(x) for x in generation]
    results = {}
    for gen in generation:
```

```
        results[gen] = {}
        agents = db_reconstruct_organisms(cur,
                 start_time, population_name, gen)
        for org in agents:
            identity = org.status['identity']
            if datafield == 'all':
                results[gen][identity] = org.status
            else:
                results[gen][identity] = \
                    org.status[datafield]
    return results

def db_get_organisms_genome(cur, start_time,
    population_name, generation='all'):
'''Analysis helper function to get entire genome of
organisms for one or more generations within a simulation
(as identified by the starting time of the simulation).

@param cur: Database cursor from connect_database()
function.
@param start_time: Starting time of current simulation in
the format of <date>-<seconds since epoch>; for example,
2013-10-11-1381480985.77.
@param population_name: Name of the population.
@param generation: Generation(s) to extract
Organism.genome. Default = 'all', which extracts
Organism.genome for all logged generations within the
specific simulation.
@type generation: list
@return: A dictionary of dictionary of genome -
{<generation count>: {<Organism identity>: <genome as list
of chromosomes>}}.'''
    if generation == 'all':
        generation = db_list_generations(cur, start_time,
                                              'organisms')
    else:
        generation = [str(x) for x in generation]
    results = {}
    for gen in generation:
        results[gen] = {}
        agents = db_reconstruct_organisms(cur,
                 start_time, population_name, gen)
        for org in agents:
            identity = org.status['identity']
            results[gen][identity] = org.genome
    return results

def db_get_organisms_chromosome_sequences(cur,
```

```
    start_time, population_name, generation='all'):
'''Analysis helper function to get chromosomal sequences
of entire genome of organisms for one or more generations
within a simulation (as identified by the starting time of
the simulation).

@param cur: Database cursor from connect_database()
function.
@param start_time: Starting time of current simulation in
the format of <date>-<seconds since epoch>; for example,
2013-10-11-1381480985.77.
@param population_name: Name of the population.
@param generation: Generation(s) to extract
Organism.genome. Default = 'all', which extracts
Organism.genome for all logged generations within the
specific simulation.
@type generation: list
@return: A dictionary of dictionary of chromosomal
sequences - {<generation count>: {<Organism identity>:
<list (all chromosomes) of list (individual chromosome) of
chromosomal sequence>}}.'''
    genome_dict = db_get_organisms_genome(cur,
                  start_time, population_name,
                  generation)
    results = {}
    for gen in list(genome_dict.keys()):
        results[gen] = {}
        for identity in list(genome_dict[gen].keys()):
            results[gen][identity] = \
                [chromosome.sequence
                 for chromosome in
                 genome_dict[gen][identity]]
    return results

def db_reconstruct_simulation_parameters(cur,
                                         start_time):
'''Function to reconstruct simulation parameters
dictionary of a simulation (as identified by the starting
time of the simulation).

@param cur: Database cursor from connect_database()
function.
@param start_time: Starting time of current simulation in
the format of <date>-<seconds since epoch>; for example,
2013-10-11-1381480985.77.
@return: Simulation parameters dictionary containing the
following keys:
    - simulation_name
```

```
    - population_names
    - population_locations
    - deployment_code
    - chromosome_bases
    - background_mutation
    - additional_mutation
    - mutation_type
    - chromosome_size
    - genome_size
    - max_tape_length
    - clean_cell
    - interpret_chromosome
    - max_codon
    - population_size
    - eco_cell_capacity
    - world_x
    - world_y
    - world_z
    - goal
    - maximum_generations
    - fossilized_ratio
    - fossilized_frequency
    - print_frequency
    - ragaraja_version
    - ragaraja_instructions
    - eco_buried_frequency
    - database_file
    - database_logging_frequency'''
    parameters = {}
    for key in ('simulation_name', 'population_names',
                'population_locations',
                'deployment_code', 'chromosome_bases',
                'background_mutation',
                'additional_mutation', 'mutation_type',
                'chromosome_size', 'genome_size',
                'max_tape_length', 'clean_cell',
                'interpret_chromosome', 'max_codon',
                'population_size', 'eco_cell_capacity',
                'world_x', 'world_y', 'world_z', 'goal',
                'maximum_generations',
                'fossilized_ratio',
                'fossilized_frequency',
                'print_frequency', 'ragaraja_version',
                'ragaraja_instructions',
                'eco_buried_frequency', 'database_file',
                'database_logging_frequency'):
        parameters[key] = None
    start_time = str(start_time)
```

```
cur.execute("select distinct simulation_name from \
    parameters where start_time = '%s'" % start_time)
parameters["simulation_name"] = \
    str(cur.fetchone()[0])
cur.execute("select key, value from parameters \
    where start_time = '%s'" % start_time)
for r in cur.fetchall():
    if str(r[0]) == 'population_names':
        value = str(r[1]).split('|')
        exec("parameters['population_names'] = %s" %\
             str(value))
    elif str(r[0]) == 'population_locations':
        exec("parameters['population_locations'] =\
             %s" % str(r[1]))
    elif str(r[0]) == 'deployment_code':
        parameters['deployment_code'] = int(r[1])
    elif str(r[0]) == 'chromosome_bases':
        value = str(r[1]).split('|')
        exec("parameters['chromosome_bases'] = %s" \
             % str(value))
    elif str(r[0]) == 'background_mutation':
        parameters['background_mutation'] = \
            float(r[1])
    elif str(r[0]) == 'additional_mutation':
        parameters['additional_mutation'] = \
            float(r[1])
    elif str(r[0]) == 'mutation_type':
        parameters['mutation_type'] = str(r[1])
    elif str(r[0]) == 'chromosome_size':
        parameters['chromosome_size'] = int(r[1])
    elif str(r[0]) == 'genome_size':
        parameters['genome_size'] = int(r[1])
    elif str(r[0]) == 'max_tape_length':
        parameters['max_tape_length'] = int(r[1])
    elif str(r[0]) == 'clean_cell':
        exec("parameters['clean_cell'] = %s" % \
             str(r[1]))
    elif str(r[0]) == 'interpret_chromosome':
        exec("parameters['interpret_chromosome'] = \
             %s" % str(r[1]))
    elif str(r[0]) == 'max_codon':
        parameters['max_codon'] = int(r[1])
    elif str(r[0]) == 'population_size':
        parameters['population_size'] = int(r[1])
    elif str(r[0]) == 'eco_cell_capacity':
        parameters['eco_cell_capacity'] = int(r[1])
    elif str(r[0]) == 'world_x':
        parameters['world_x'] = int(r[1])
```

```
        elif str(r[0]) == 'world_y':
            parameters['world_y'] = int(r[1])
        elif str(r[0]) == 'world_z':
            parameters['world_z'] = int(r[1])
        elif str(r[0]) == 'goal':
            try: parameters['goal'] = float(r[1])
            except ValueError:
               exec("parameters['goal'] = %s" % \
                    str(r[1]))
        elif str(r[0]) == 'maximum_generations':
            parameters['maximum_generations'] = int(r[1])
        elif str(r[0]) == 'fossilized_ratio':
            parameters['fossilized_ratio'] = float(r[1])
        elif str(r[0]) == 'fossilized_frequency':
            parameters['fossilized_frequency'] = \
                int(r[1])
        elif str(r[0]) == 'print_frequency':
            parameters['print_frequency'] = int(r[1])
        elif str(r[0]) == 'ragaraja_version':
            version = str(r[1])
            if version == '0.1':
                parameters['ragaraja_version'] = 0.1
            else:
                parameters['ragaraja_version'] = \
                    int(r[1])
        elif str(r[0]) == 'ragaraja_instructions':
            value = str(r[1]).split('|')
            exec("parameters['ragaraja_instructions'] = \
                 %s" % str(value))
        elif str(r[0]) == 'eco_buried_frequency':
            parameters['eco_buried_frequency'] = \
                int(r[1])
        elif str(r[0]) == 'database_file':
            parameters['database_file'] = str(r[1])
        elif str(r[0]) == 'database_logging_frequency':
            parameters['database_logging_frequency'] = \
                int(r[1])
    return parameters

def db_reconstruct_world(cur, start_time, generation):
'''Function to reconstruct the world object of a
simulation (as identified by the starting time of the
simulation) at a specific generation.

@param cur: Database cursor from connect_database()
function.
```

```
@param start_time: Starting time of current simulation in
the format of <date>-<seconds since epoch>; for example,
2013-10-11-1381480985.77.
@param generation: Current number of generations
simulated.
@return: dose_world.World object'''
    import dose_world
    eco_cell = {'local_input': [], 'local_output': [],
                'temporary_input': [],
                'temporary_output': [],
                'organisms': 0}
    cur.execute("select max(x), max(y), max(z) from \
        world where start_time = '%s'" % start_time)
    coordinates = cur.fetchone()
    world_x = int(coordinates[0]) + 1
    world_y = int(coordinates[1]) + 1
    world_z = int(coordinates[2]) + 1
    ecosystem = {}
    for x in range(world_x):
        ecosystem[x] = {}
        for y in range(world_y):
            ecosystem[x][y] = {}
            for z in range(world_z):
                ecosystem[x][y][z] = {}
                for key in ('local_input',
                            'local_output',
                            'temporary_input',
                            'temporary_output',
                            'organisms'):
                    ecosystem[x][y][z][key] = None
    start_time = str(start_time)
    generation = str(generation)
    # query plan: SCAN TABLE world
    cur.execute("select x, y, z, key, value from \
        world where start_time = '%s' and \
        generation = '%s'" % (start_time, generation))
    for r in cur.fetchall():
        x = int(r[0])
        y = int(r[1])
        z = int(r[2])
        key = str(r[3])
        value = str(r[4])
        exec("ecosystem[%i][%i][%i]['%s'] = %s" % \
            (x, y, z, key, value))
    World = dose_world.World(1, 1, 1)
    World.ecosystem = ecosystem
    return World
```

```
def db_reconstruct_organisms(cur, start_time,
    population_name, generation):
'''Function to reconstruct a list of organisms
(genetic.Organism objects) of a population within a
simulation (as identified by the starting time of the
simulation) at a specific generation.

@param cur: Database cursor from connect_database()
function.
@param start_time: Starting time of current simulation in
the format of <date>-<seconds since epoch>; for example,
2013-10-11-1381480985.77.
@param population_name: Name of the population
@param generation: Current number of generations
simulated.
@return: A list of Organisms (genetic.Organism objects)
'''
    import genetic as g
    start_time = str(start_time)
    population_name = str(population_name)
    generation = str(generation)
    # query plan: SEARCH TABLE organisms USING INDEX
    #             organisms_index2 (generation=?)
    cur.execute("select distinct org_name from \
        organisms where start_time = '%s' and \
        pop_name = '%s' and generation = '%s'" % \
        (start_time, population_name, generation))
    names = [x[0] for x in cur.fetchall()]
    agents = [0] * len(names)
    for i in range(len(names)):
        org_name = str(names[i])
        org = g.Organism()
        org.genome = []
        org.status['identity'] = org_name
        # query plan: SEARCH TABLE organisms USING INDEX
        #             organisms_index2 (generation=? AND
        #             org_name=?)
        cur.execute("select key, value from organisms \
            where generation = '%s' and org_name = '%s' \
            and start_time = '%s' and pop_name = '%s'" %\
        (generation, org_name, start_time,
         population_name))
        for r in cur.fetchall():
            key = str(r[0])
            value = str(r[1])
            if key == 'alive':
                exec("org.status['alive'] = %s" % \
                     str(value))
```

```
elif key == 'vitality':
    org.status['vitality'] = float(value)
elif key == 'parents':
    if value == '': value = '[]'
    else: value = value.split('|')
    exec("org.status['parents'] = %s" % \
         str(value))
elif key == 'age':
    org.status['age'] = float(value)
elif key == 'gender':
    exec("org.status['gender'] = %s" % \
         str(value))
elif key == 'lifespan':
    org.status['lifespan'] = float(value)
elif key == 'fitness':
    exec("f = '%s'" % str(value))
    if type(f) == type(1) or \
       type(f) == type(1.0):
        org.status['fitness'] = float(value)
    else:
        exec("org.status['fitness'] = %s" % \
             str(value))
elif key == 'blood':
    if value == '': value = '[]'
    else: value = value.split('|')
    exec("org.status['blood'] = %s" % \
         str(value))
elif key == 'deme':
    org.status['deme'] = str(value)
elif key == 'location':
    value = tuple([int(x)
                  for x in value.split('|')])
    exec("org.status['location'] = %s" % \
         str(value))
elif key == 'death':
    exec("org.status['death'] = %s" % \
         str(value))
elif key.startswith('chromosome'):
    chr_position = key.split('_')[1]
    sequence = [str(x) for x in str(value)]
    cur.execute("select value from \
        parameters where \
        key='background_mutation' and \
        start_time='%s'" % start_time)
    background_mutation = \
        float(cur.fetchone()[0])
    cur.execute("select value from \
        parameters where \
```

```
                    key='chromosome_bases' and \
                    start_time='%s'" % start_time)
                bases = str(cur.fetchone()[0]).split('|')
                exec("chromosome_bases = %s" % bases)
                chromosome = g.Chromosome(sequence,
                             chromosome_bases,
                             background_mutation)
                org.genome.append(chromosome)
            else:
                exec("org.status['%s'] = %s" % \
                    (str(key), str(value)))
        agents[i] = org
    return agents

def db_reconstruct_population(cur, start_time,
    population_name, generation):
'''Function to reconstruct a population within a
simulation (as identified by the starting time of the
simulation) at a specific generation.

@param cur: Database cursor from connect_database()
function.
@param start_time: Starting time of current simulation in
the format of <date>-<seconds since epoch>; for example,
2013-10-11-1381480985.77.
@param population_name: Name of the population
@param generation: Current number of generations
simulated.
@return: genetic.Population object'''
    import genetic
    agents = db_reconstruct_organisms(cur, start_time,
                                      population_name,
                                      generation)
    cur.execute("select value from parameters where \
          key='goal' and start_time='%s'" % start_time)
    g = cur.fetchone()[0]
    try: goal = float(g)
    except ValueError: exec("goal = %s" % str(g))
    return genetic.Population(goal, 1e24, agents)
```

File Name: simulation_calls.py

```
'''File containing support functions for running a
simulation. The functions in this file is not for public
use; all functions in this file are private functions.
```

```
Date created: 10th October 2013
'''
import random, inspect, os
import os.path
from datetime import datetime
from time import time
from copy import deepcopy
from shutil import copyfile

try:
    import cPickle as pickle
except ImportError:
    import pickle

import dose_world
import genetic
import ragaraja, register_machine

from database_calls import connect_database
from database_calls import db_log_simulation_parameters
from database_calls import db_report

def simulation_core(sim_functions, sim_parameters,
                    Populations, World):
'''Sequential ecological cell DOSE simulator.

Performs the following operations:
     - Creating simulation file directory for results text
     file, population freeze, and world burial storage
     - Generate a simulation start time to identify the
     current simulation
    - Define active Ragaraja instructions
    - Connecting to logging database (if needed)
     - Writing simulation parameters into results text
     file
     - Initialize World and Population
     - Deploy population(s) onto the world
     - Run the simulation and recording the results
     - Writing final results into results text file
     - Close logging database (if used)
     - Copy simulation script into simulation file
     directory

@param sim_functions: implemented simulation functions
(see sose.dose_functions)
@param sim_parameters: simulation parameters dictionary
(see Examples)
```

```
@param Populations: dictionary of population objects
@param World: dose_world.World object'''
    time_start = \
        '-'.join([str(datetime.utcnow()).split(' ')[0],
                  str(time())])
    print('Creating simulation file directories...')
    directory = \
        '_'.join([sim_parameters["simulation_name"],
                  time_start])
    directory = os.sep.join([os.getcwd(), 'Simulations',
                             directory])
    directory = directory + os.sep
    if not os.path.exists(directory):
        os.makedirs(directory)
    print('Adding simulation directory to simulation \
          parameters...')
    sim_parameters["directory"] = directory
    print('Adding starting time to simulation \
          parameters...')
    sim_parameters["starting_time"] = time_start
    sim_functions = sim_functions()
    if sim_parameters["ragaraja_version"] == 0:
        print('Activating ragaraja version: 0...')
        ragaraja.activate_version(
            sim_parameters["ragaraja_version"],
            sim_parameters["ragaraja_instructions"])
    else:
        print('Activating ragaraja version: ' + \
            str(sim_parameters["ragaraja_version"]) + \
            '...')
        ragaraja.activate_version(
            sim_parameters["ragaraja_version"])
    if "database_file" in sim_parameters and \
        "database_logging_frequency" in sim_parameters:
        print('Connecting to database file: ' + \
              sim_parameters["database_file"] + '...')
        (con, cur) = connect_database(None,
                                      sim_parameters)
        print('Logging simulation parameters to \
              database file...')
        (con, cur) = db_log_simulation_parameters(con,
                     cur, sim_parameters)
    for pop_name in Populations:
        print('\nPreparing population: ' + pop_name + ' \
              for simulation...')
        if 'sim_folder' in sim_parameters or \
            'database_source' in sim_parameters:
            print('Calculating final generation \
```

```
            count...')
        max = sim_parameters["rev_start"] \
              [sim_parameters["population_names"].
                  index(pop_name)] + \
               sim_parameters["extend_gen"]
        print('Updating generation count from \
              previous simulation...')
        generation_count = \
            sim_parameters["rev_start"][0]
    else:
        print('Deploying population in World \
              entity...')
        if sim_parameters["deployment_code"] == 0:
            print('Executing user defined \
                  deployment scheme...')
            deploy_0(sim_parameters, Populations,
                     pop_name, World)
        elif sim_parameters["deployment_code"] == 1:
            print('Executing deployment code 1: \
                  Single eco-cell deployment...')
            deploy_1(sim_parameters, Populations,
            pop_name, World)
        elif sim_parameters["deployment_code"] == 2:
            print('Executing deployment code 2: \
                  Random eco-cell deployment...')
            deploy_2(sim_parameters, Populations,
                     pop_name, World)
        elif sim_parameters["deployment_code"] == 3:
            print('Executing deployment code 3: \
                  Even eco-cell deployment...')
            deploy_3(sim_parameters, Populations,
                     pop_name, World)
        elif sim_parameters["deployment_code"] == 4:
            print('Executing deployment code 4: \
                  Centralized eco-cell \
                  deployment...')
            deploy_4(sim_parameters, Populations,
                     pop_name, World)
        print('Adding maximum generations to \
              simulation parameters...')
        max = sim_parameters["maximum_generations"]
        generation_count = 0
    print('Writing simulation parameters into txt \
          file report...')
    write_parameters(sim_parameters, pop_name)
    print('Updating generation count...')
    Populations[pop_name].generation = \
        generation_count
```

```
    print('\nSimulation preparation complete...')
    while generation_count < max:
        generation_count = generation_count + 1
        sim_functions.ecoregulate(World)
        eco_cell_iterator(World, sim_parameters,
                          sim_functions.update_ecology)
        eco_cell_iterator(World, sim_parameters,
                          sim_functions.update_local)
        eco_cell_iterator(World, sim_parameters,
                          sim_functions.report)
        bury_world(sim_parameters, World,
                   generation_count)
        for pop_name in Populations:
            if sim_parameters["interpret_chromosome"]:
                interpret_chromosome(sim_parameters,
                                     Populations,
                                     pop_name, World)
            report_generation(sim_parameters,
                              Populations, pop_name,
                              sim_functions,
                              generation_count)
            sim_functions.organism_movement(Populations,
                                            pop_name, World)
            sim_functions.organism_location(Populations,
                                            pop_name, World)
        if "database_file" in sim_parameters and \
            "database_logging_frequency" in \
                sim_parameters and \
            generation_count % \
                int(sim_parameters
                ["database_logging_frequency"]) == 0:
                (con, cur) = db_report(con, cur,
                              sim_functions,
                              sim_parameters["starting_time"],
                              Populations, World,
                              generation_count)
        print('Generation ' + str(generation_count) + ' \
             complete...')
    print('\nClosing simulation results...')
    for pop_name in Populations:
        close_results(sim_parameters, pop_name)
    if "database_file" in sim_parameters and \
        "database_logging_frequency" in sim_parameters:
        print('Committing logged data into database \
             file...')
        con.commit()
        print('Terminating database connection...')
        con.close()
```

```
    print('Copying simulation file script to simulation \
         results directory...')
    sim_script_basename = \
        os.path.basename(inspect.stack()[2][1])
    copyfile(inspect.stack()[2][1],
             os.path.join(sim_parameters['directory'],
             sim_script_basename))
    print('\nSimulation ended...')

def coordinates(location):
'''Helper function to transpose ecological cell into a
tuple.

@param location: location of ecological cell as 3-element
iterable data type
@return: location of ecological cell as (x,y,z)'''
    x = location[0]
    y = location[1]
    z = location[2]
    return (x,y,z)

def adjacent_cells(sim_parameters, location):
'''Function to get a list of adjacent ecological cells
from a given location.

@param sim_parameters: simulation parameters dictionary
(see Examples)
@param location: location of ecological cell as (x,y,z)
@return: list of locations of adjacent cells.'''
    trashbin = []
    temp_cells = []
    world_size = [sim_parameters["world_x"],
                  sim_parameters["world_y"],
                  sim_parameters["world_z"]]
    for i in range(3):
        new_location = [spot for spot in location]
        new_location[0] = new_location[0] + 1
        temp_cells.append(new_location)
        new_location = [spot for spot in location]
        new_location[0] = new_location[0] - 1
        temp_cells.append(new_location)
    for i in range(2):
        new_location = [spot for spot in location]
        new_location[1] = new_location[1] - 1
        temp_cells.append(new_location)
    for i in range(0,4,3):
        temp_cells[i][1] = temp_cells[i][1] + 1
```

```
            temp_cells[i+1][1] = temp_cells[i+1][1] - 1
        temp_cells[-1][1] = temp_cells[-1][1] + 2
        for i in range(8):
            for x in range(2):
                if temp_cells[i][x] >= world_size[x] or \
                    temp_cells[i][x] < 0:
                    if temp_cells[i] not in trashbin:
                        trashbin.append(temp_cells[i])
        for location in trashbin:
            temp_cells.remove(location)
        return [tuple(location) for location in temp_cells]

def spawn_populations(sim_parameters):
'''Initializing starting population(s) for a simulation.
Each organism in each population at this stage will be
genetic clones of each other.

@param sim_parameters: simulation parameters dictionary
(see Examples)
@return: dictionary of population objects with population
name as key'''
    temp_Populations = {}
    print(' - Accessing population names...')
    for pop_name in sim_parameters["population_names"]:
        print(' - Constructing population: ' + \
              pop_name + '...')
        temp_Populations[pop_name] = \
           genetic.population_constructor(sim_parameters)
        print(' - Updating organism identity and deme \
              status...')
        for individual in temp_Populations
            [pop_name].agents:
            individual.generate_name()
            individual.status['deme'] = pop_name
    return temp_Populations

def eco_cell_iterator(World, sim_parameters, function):
'''Generic caller to call any function to be executed in
each ecological cell in sequence.

@param World: dose_world.World object
@param sim_parameters: simulation parameters dictionary
(see Examples)
@param function: function to be executed
@return: none'''
    for x in range(sim_parameters["world_x"]):
        for y in range(sim_parameters["world_y"]):
            for z in range(sim_parameters["world_z"]):
```

```
                if len(inspect.getargspec(function)[0]) \
                    == 5:
                    function(World, x, y, z)
                else:
                    function(World)
def deploy_0(sim_parameters, Populations,
             pop_name, World):
'''Organism deployment scheme 0 - User defined deployment
scheme. This is called when "deployment_code" in
simulation parameters = 0. User will have to provide a
deployment scheme/function as "deployment_scheme" in
simulation parameters.

@param sim_parameters: simulation parameters dictionary
(see Examples)
@param Populations: dictionary of population objects
@param pop_name: population name
@param World: dose_world.World object
@return: none.'''
    sim_parameters["deployment_scheme"](Populations,
                                        pop_name, World)

def deploy_1(sim_parameters, Populations,
             pop_name, World):
'''Organism deployment scheme 1 - Single ecological cell
deployment scheme. This is called when "deployment_code"
in simulation parameters = 1. In this scheme, all
organisms in the specified population (specified by
population name) will be deployed in the ecological cell
specified in "population_locations" of simulation
parameters.

@param sim_parameters: simulation parameters dictionary
(see Examples)
@param Populations: dictionary of population objects
@param pop_name: population name
@param World: dose_world.World object
@return: none.'''
    position = \
       sim_parameters["population_names"].index(pop_name)
    locations = [location
        for location in
        sim_parameters["population_locations"][position]]
    (x,y,z) = coordinates(locations[0])
    World.ecosystem[x][y][z]['organisms'] = \
        sim_parameters["population_size"]
    for individual in Populations[pop_name].agents:
        individual.status['location'] = locations[0]
```

```
def deploy_2(sim_parameters, Populations,
             pop_name, World):
'''Organism deployment scheme 2 - Random deployment
scheme. This is called when "deployment_code" in
simulation parameters = 2. In this scheme, all organisms
in the specified population (specified by population
name) will be randomly deployed across the list of
ecological cells specified as "population_locations" of
simulation parameters.

@param sim_parameters: simulation parameters dictionary
(see Examples)
@param Populations: dictionary of population objects
@param pop_name: population name
@param World: dose_world.World object
@return: none'''
    position = \
       sim_parameters["population_names"].index(pop_name)
    locations = [location
        for location in
        sim_parameters["population_locations"][position]]
    for individual in Populations[pop_name].agents:
            location = random.choice(locations)
            (x,y,z) = coordinates(location)
            while World.ecosystem[x][y][z]['organisms'] \
                >= sim_parameters["eco_cell_capacity"]:
                location = random.choice(locations)
                (x,y,z) = coordinates(location)
            World.ecosystem[x][y][z]['organisms'] = \
                World.ecosystem[x][y][z]['organisms'] + 1
            individual.status['location'] = location

def deploy_3(sim_parameters, Populations,
             pop_name, World):
'''Organism deployment scheme 3 - Even deployment scheme.
This is called when "deployment_code" in simulation
parameters = 3. In this scheme, all organisms in the
specified population (specified by population name) will
be evenly deployed (as much as possible) across the list
of ecological cells specified as "population_locations" of
simulation parameters.

@param sim_parameters: simulation parameters dictionary
(see Examples)
@param Populations: dictionary of population objects
@param pop_name: population name
@param World: dose_world.World object
@return: none'''
```

```
    position = \
    sim_parameters["population_names"].index(pop_name)
    locations = [location
        for location in
        sim_parameters["population_locations"][position]]
    iterator = 0
    for i in range(sim_parameters["population_size"]):
        individual = Populations[pop_name].agents[i]
        location = locations[iterator]
        (x,y,z) = coordinates(location)
        World.ecosystem[x][y][z]['organisms'] = \
            World.ecosystem[x][y][z]['organisms'] + 1
        individual.status['location'] = location
        iterator = iterator + 1
        if iterator == len(locations):
            iterator = 0

def deploy_4(sim_parameters, Populations,
             pop_name, World):
'''Organism deployment scheme 4 - Centralized deployment
scheme. This is called when "deployment_code" in
simulation parameters = 4. In this scheme, all organisms
in the specified population (specified by population name)
will be deployed onto the ecological cell specified
in "population_locations" of simulation parameters, up to
the organism capacity of the ecological cell (specified in
"eco_cell_capacity" of simulation parameters). In event
that the specified deployment locations are filled,
undeployed organisms will be randomly deployed onto
adjacent cells.

@param sim_parameters: simulation parameters dictionary
(see Examples)
@param Populations: dictionary of population objects
@param pop_name: population name
@param World: dose_world.World object
@return: none'''
    position = \
    sim_parameters["population_names"].index(pop_name)
    locations = [location
        for location in
        sim_parameters["population_locations"][position]]
    adj_cells = adjacent_cells(sim_parameters,
                               locations[0])
    for group in range((sim_parameters
        ["population_size"] / \
        sim_parameters["eco_cell_capacity"]) + 1):
```

```
        start = sim_parameters["eco_cell_capacity"] * \
                group
        end = start + sim_parameters["eco_cell_capacity"]
        for x in range(start, end):
            if x == sim_parameters["population_size"]:
                break
            individual = Populations[pop_name].agents[x]
            if x > \
               (sim_parameters["eco_cell_capacity"] - 1):
                location = random.choice(adj_cells)
                (x,y,z) = coordinates(location)
                while World.ecosystem[x][y][z]
                    ['organisms'] > \
                    sim_parameters["eco_cell_capacity"]:
                    location = random.choice(adj_cells)
                    (x,y,z) = \
                    coordinates(random.choice(adj_cells))
            (x,y,z) = coordinates(location)
            World.ecosystem[x][y][z]['organisms'] = \
                World.ecosystem[x][y][z]['organisms'] + 1
            individual.status['location'] = location

def interpret_chromosome(sim_parameters, Populations,
                         pop_name, World):
'''Function to call Ragaraja interpreter to express /
execute the genome for each organism in a population. The
Turing tape (array) after execution will be logged as
"blood" of each organism. Ragaraja interpreter will use
the temporary_input and temporary_output lists from each
ecological cell as the input data and output data
respectively for genome execution - this will resemble
consumption of environmental resources and replenishing of
environmental resources or dumping of wastes respectively.

@param sim_parameters: simulation parameters dictionary
(see Examples)
@param Populations: dictionary of population objects
@param pop_name: population name
@param World: dose_world.World object
@return: none'''
    array = [0] * sim_parameters["max_tape_length"]
    for i in range(len(Populations[pop_name].agents)):
        individual = Populations[pop_name].agents[i]
        location = individual.status['location']
        (x,y,z) = coordinates(location)
        if sim_parameters["clean_cell"]:
            array = [0] * \
```

```
                    sim_parameters["max_tape_length"]
        else:
            array = \
                Populations[pop_name].
                agents[i].status['blood']
            if array == None:
                array = [0] * \
                    sim_parameters["max_tape_length"]
        for chromosome_count in
            range(len(individual.genome)):
            inputdata = \
                World.ecosystem[x][y][z]['local_input']
            output = \
                World.ecosystem[x][y][z]['local_output']
            source = \
                ''.join(individual.genome
                        [chromosome_count].sequence)
            array = \
                Populations[pop_name].
                agents[i].status['blood']
            try:
                (array, apointer, inputdata,
                 output, source, spointer) = \
                register_machine.interpret(source,
                ragaraja.ragaraja, 3, inputdata,
                array, sim_parameters["max_tape_length"],
                sim_parameters["max_codon"])
            except Exception as e:
                error_msg = \
                    '|'.join(['Error at Chromosome_' + \
                    str(chromosome_count), str(e)])
                Populations[pop_name].agents[i]. \
                    status['chromosome_error'] = \
                       error_msg
                Populations[pop_name].agents[i].
                    status['blood'] = array
            Populations[pop_name].agents[i].
                status['blood'] = array
            World.ecosystem[x][y][z]['temporary_input'] \
                = inputdata
            World.ecosystem[x][y][z]['temporary_output']\
                = output

def step(Populations, pop_name, sim_functions):
'''Performs a generational step for a population
    - Prepopulation control
    - Mutations
    - Before mating fitness measurement
```

```
    - Mating
    - Postpopulation control
    - Generational events
    - After mating fitness measurement
    - Generate a textual report for the current
    generation

@param Populations: dictionary of population objects
@param pop_name: population name
@param sim_functions: implemented simulation functions
(see dose.dose_functions)
@return: report as a string'''
    if Populations[pop_name].generation > 0:
        sim_functions.prepopulation_control(Populations,
                                             pop_name)
    for organism in Populations[pop_name].agents:
        sim_functions.mutation_scheme(organism)
    sim_functions.fitness(Populations, pop_name)
    sim_functions.mating(Populations, pop_name)
    sim_functions.postpopulation_control(Populations,
                                          pop_name)
    sim_functions.generation_events(Populations,
                                     pop_name)
    Populations[pop_name].generation = \
        Populations[pop_name].generation + 1
    sim_functions.fitness(Populations, pop_name)
    return sim_functions.population_report(Populations,
                                            pop_name)

def report_generation(sim_parameters, Populations,
    pop_name, sim_functions, generation_count):
'''Performs a generational step (using step function) for
a population and writes out the resulting report into
results text file.

@param sim_parameters: simulation parameters dictionary
(see Examples)
@param Populations: dictionary of population objects
@param pop_name: population name
@param sim_functions: implemented simulation functions
(see dose.dose_functions)
@param generation_count: current generation count for
reporting
@return: none'''
    report = step(Populations, pop_name, sim_functions)
    if generation_count % \
        int(sim_parameters["fossilized_frequency"]) == 0:
        file = '%s%s_%s_' % (sim_parameters["directory"],
```

```
            sim_parameters["simulation_name"], pop_name)
        freeze_population(file,
                sim_parameters["fossilized_ratio"],
                Populations, pop_name)
    if generation_count % \
        int(sim_parameters["print_frequency"]) == 0:
        f = open(('%s%s_%s.result.txt' % \
          (sim_parameters["directory"],
           sim_parameters["simulation_name"],
           pop_name)), 'a')
        dtstamp = str(datetime.utcnow())
        f.write('\n'.join(['\n' + dtstamp,
                           'GENERATION: ' + \
                           str(generation_count),
                           str(report)]))
        f.write('\n')
        f.close

def bury_world(sim_parameters, World, generation_count):
'''Function to bury entire world into a file.

@param sim_parameters: simulation parameters dictionary
(see Examples)
@param World: dose_world.World object
@param generation_count: current generation count for file
name generation'''
    if generation_count % \
    int (sim_parameters["eco_buried_frequency"]) == 0:
       filename = '%s%s_gen%s.eco' % \
                  (sim_parameters["directory"],
                   sim_parameters["simulation_name"],
                   str(generation_count))
       f = open(filename, 'wb')
       pickle.dump(World, f)
       f.close()

def excavate_world(eco_file):
'''Excavate buried world from file.

@param eco_file: buried world file generated by bury_world
function.
@return: excavated dose_world.World object'''
    f = open(eco_file, 'rb')
    return pickle.load(f)

def freeze_population(file, proportion,
                      Populations, pop_name):
```

```
'''Function to freeze part or whole of the population into
a file. If the number of organisms is less than 101, the
entire population will be frozen.

@param file: file name prefix
@param proportion: proportion of the population to freeze
@param Populations: dictionary of population objects
@param pop_name: population name to freeze'''
    if proportion > 1.0: proportion = 1.0
    agents = Populations[pop_name].agents
    if len(agents) < 101 or \
        len(agents) * proportion < 101:
        sample = deepcopy(Populations[pop_name])
    else:
        new_agents = [agents[random.randint(0,
                      len(agents) - 1)]
                      for x in range(int(len(agents) * \
                      proportion))]
        sample = deepcopy(Populations[pop_name])
        sample.agents = new_agents
    name = ''.join([file, 'pop',
                   str(Populations[pop_name].generation),
                   '_', str(len(sample.agents)), '.gap'])
    f = open(name, 'wb')
    pickle.dump(sample, f)
    f.close()

def revive_population(gap_file):
'''Revive population from a frozen population file.

@param gap_file: frozen population file generated by
freeze_population function.
@return: revived population object'''
    f = open(gap_file, 'rb')
    return pickle.load(f)

def write_parameters(sim_parameters, pop_name):
'''Function to write simulation parameters into results
text file as header.

@param sim_parameters: simulation parameters dictionary
(see Examples)
@param pop_name: population name
@return: none'''
    f = open(('%s%s_%s.result.txt' % \
             (sim_parameters["directory"],
              sim_parameters["simulation_name"],
              pop_name)), 'a')
```

```
    f.write("""SIMULATION: %s
-------------------------------------------------------
""" % sim_parameters["simulation_name"])
    if ('database_source' in sim_parameters) or \
        ('sim_folder' in sim_parameters):
        f.write("SIMULATION REVIVAL STARTED: %s\n\n" % \
                sim_parameters["starting_time"])
    else:
        f.write("SIMULATION STARTED: %s\n\n" % \
                sim_parameters["starting_time"])
    for key in sim_parameters:
        if key not in ('deployment_scheme', 'directory',
                       'sim_folder'):
            f.write("%s : %s\n" % (key,
sim_parameters[key]))
    f.write("""\n\nREPORT
------------------------------------------------------
""")

def close_results(sim_parameters, pop_name):
'''Function to write footer of results text file.

@param sim_parameters: simulation parameters dictionary
(see Examples)
@param pop_name: population name
@return: none'''
    f = open(('%s%s_%s.result.txt' % \
             (sim_parameters["directory"],
              sim_parameters["simulation_name"],
              pop_name)), 'a')
    f.write('''
-------------------------------------------------------
SIMULATION ENDED: ''' + str(datetime.utcnow()))
    f.close()
```

File Name: genetic.py

```
"""Framework for Genetic Algorithm Applications.
Date created: 23rd February 2010

@see: Lim, JZR, Aw, ZQ, Goh, DJW, How, JA, Low, SXZ, Loo,
BZL, Ling, MHT. 2010. A genetic algorithm framework
grounded in biology. The Python Papers Source Codes 2: 6.
"""
import random, os, string
```

```
from copy import deepcopy

class Chromosome(object):
"""Representation of a linear chromosome.

@see: Lim, JZR, Aw, ZQ, Goh, DJW, How, JA, Low, SXZ, Loo,
BZL, Ling, MHT. 2010. A genetic algorithm framework
grounded in biology. The Python Papers Source Codes 2: 6.

@since: version 0.4"""
    def __init__(self, sequence, base,
                 background_mutation=0.0001):
    """Sets up a chromosome.

    @param sequence: a subscriptable object (list or
    string) representing the sequence of the chromosome.
    @param base: a subscriptable object (list or string)
    Representing allowable entities in the sequence.
    @param background_mutation: background mutation rate
    represented as the probability of number of mutations
    per base. Default = 0.0001 (0.01%).

    @since: version 0.4"""
        self.sequence = sequence
        self.base = base
        self.background_mutation = background_mutation

    def rmutate(self, type='point', rate=0.01,
                start=0, end=-1):
    """
    Random Mutation operator - to simulate random point,
    insertion, deletion, inversion, gene translocation
    and gene duplication events.

    The start and end parameters are useful for
    simulating mutational hotspots in the genome.

    @param type: type of mutation. Accepts 'point' (point
    mutation), 'insert' (insert a base), 'delete' (delete
    a base), 'invert' (invert a stretch of the
    chromosome), 'duplicate' (duplicate a stretch of the
    chromosome), 'translocate' (translocate a stretch of
    chromosome to another random position). Default =
    point.
    @param rate: probability of mutation per base above
    Background mutation rate. Default = 0.01 (1%). No
    mutation event will ever happen if (rate +
    background_mutation) is less than zero.
```

```
@param start: starting base on the sequence for
mutation. Default = 0, start of the genome.
@param end: last base on the sequence for mutation.
Default = -1, end of the genome.

@since: version 0.4"""
    if end > len(self.sequence) - 1:
        end = len(self.sequence) - 1
    if end == -1: end = len(self.sequence) - 1
    if start == end: start = 0
    length = int(end - start)
    mutation = int((self.background_mutation + rate)\
                   * length)
    while mutation > 0:
        position = int(start) + \
                   random.randrange(length - 1)
        new_base = \
          self.base[random.randrange(len(self.base))]
        if type == 'point':
            self.sequence[position] = new_base
        if type == 'delete':
            self.sequence.pop(position)
        if type == 'insert':
            self.sequence.insert(position, new_base)
        if type == 'duplicate':
            end_pos = random.randrange(position + 1,
                                       end)
            fragment = \
                self.sequence[position:end_pos]
            for i in range(len(fragment)):
                self.sequence.insert(end_pos + i,
                                     fragment[i])
        if type == 'invert':
            end_pos = random.randrange(position + 1,
                                       end)
            fragment = [self.sequence.pop(position)
                for i in range(end_pos - position)]
            fragment.reverse()
            for base in fragment:
                self.sequence.insert(position, base)
        if type == 'translocate':
            end_pos = random.randrange(position + 1,
                                       end)
            fragment = [self.sequence.pop(position)
                for i in range(end_pos - position)]
            insertion_point = random.randint(0,
                              len(self.sequence))
            for i in range(len(fragment)):
```

```
                self.sequence.insert(insertion_point
                                     + i, fragment[i])
        mutation = mutation - 1

    def kmutate(self, type='point',
                start=0, end=0, sequence=None, tpos=0):
    """Known Mutation operator - to simulate a known
    point, insertion, deletion, inversion, gene
    translocation or gene duplication event.

    The required parameters will be determined by the
    type of mutation:
        - point requires
        - start (the position of the base to mutate)
        - sequence (the new base)
        - delete requires
        - start (the position of the base to delete)
        - insert requires
        - start (the position of the base to begin
        insertion)
        - sequence (list of sequence or a base to insert)
        - invert requires
        - start (the position of the base to start
        inversion)
        - end (the position of the base to end inversion)
        - duplicate requires
        - start (the position of the starting base to
        duplicate)
        - end (the position of the last base to
        duplicate)
        - translocate requires
        - start (the position of the base to start
        translocation)
        - end (the position of the base to end
        translocation)
        - tpos (the position of the base insert the
        translocated sequence)

    @param type: type of mutation. Accepts 'point' (point
    mutation), 'insert' (insert a base), 'delete' (delete
    a base), 'invert' (invert a stretch of the
    chromosome), 'duplicate' (duplicate a stretch of the
    chromosome), 'translocate' (translocate a stretch of
    chromosome to another random position). Default =
    point.
    @param start: starting base on the chromosome for
    mutation. Default = 0, start of the genome.
    @param end: last base on the chromosome for mutation.
```

```
    Default = 0.
    @param sequence: list of bases to change to (point
    mutation) or to insert (insertion mutation)
    @param tpos: position of the chromosome to insert the
    Translocated sequence.

    @since: version 0.4"""
        if type == 'point':
            self.sequence[start] = sequence
        if type == 'delete':
            self.sequence.pop(start)
        if type == 'insert':
            for base in list(sequence):
                self.sequence.insert(start, base)
        if type == 'duplicate':
            fragment = self.sequence[start:end + 1]
            for i in range(len(fragment)):
                self.sequence.insert(end + i,
                                     fragment[i])
        if type == 'invert':
            for base in self.sequence[start:end]:
                self.sequence.insert(start, base)
        if type == 'translocate':
            fragment = [self.sequence.pop(start)
                        for i in range(end - start)]
            for i in range(len(fragment)):
                self.sequence.insert(tpos + i,
                                     fragment[i])

    def replicate(self):
    """Replicates (deep copy) the chromosome.

    @return: a copy of current chromosome.

    @since: version 0.4"""
        return deepcopy(self)


class Organism(object):
"""An organism represented by a list of chromosomes and a
status table. This class should almost never be
instantiated on its own but acts as an ancestor class as
some functions need to be over-ridden in the inherited
class for the desired use.

Each organism is identifiable by a randomly generated 32-
character 'name' as Organism.identity.
```

```
Methods to be over-ridden in the inherited class or
substituted are
    - fitness
    - mutation_scheme

Pre-defined status are
    1. alive - is the organism alive? True or False.
    2. vitality - percentage of maximum vitality.
    3. age - the current age of the organism.
    4. lifespan - pre-defined maximum lifespan to be set
    based on scenario.
    5. fitness - how fit the organism is? Set to maximum
    fitness.
    6. death - reason of death (as death code).
    7. parents - identity of parents.
    8. gender - gender of organism.
    9. blood - result of genomic interpretation or
    expression by Rajaraga interpreter.
    10. identity - 32-character randomly generated name.
    11. deme - defined as a sub-population or local
    population.
    12. location - location of the organism within the
    ecosystem.


List of defined death codes
    1. death01 - zero vitality
    2. death02 - maximum age reached
    3. death03 - unknown death cause

@see: Lim, JZR, Aw, ZQ, Goh, DJW, How, JA, Low, SXZ, Loo,
BZL, Ling, MHT. 2010. A genetic algorithm framework
grounded in biology. The Python Papers Source Codes 2: 6.

@since: version 0.4"""

    def __init__(self, genome='dummy',
                 mutation_type='point',
                 additional_mutation_rate=0.01,
                 gender=None):
    """Sets up a new organism with default status (age =
    0, vitality = 100, lifespan = 100, fitness = 100,
    alive = True)

    @param genome: list of chromosomes to inherit.
    Default = 'dummy'. This creates a 'dummy' chromosome
    which is basically a one-base chromosome - this is
```

```
    for applications which does not utilize the
    chromosome.
    @param mutation_type: type of mutation. Accepts
    'point' (point mutation), 'insert' (insert a base),
    'delete' (delete a base), 'invert' (invert a stretch
    of the chromosome), 'duplicate' (duplicate a stretch
    of the chromosome), 'translocate' (translocate a
    stretch of chromosome to another random position).
    Default = point.
    @param additional_mutation_rate: probability of
    mutation per base above background mutation rate.
    Default = 0.01 (1%). No mutation event will ever
    happen if (rate + background_mutation) is less than
    zero.
    @param gender: establishes the gender of the organism
    which may be used for mating routines.

    @since: version 0.4"""
        self.status = {'alive': True,
                       # is the organism alive?
                       'vitality': 100.0,
                       # % of vitality
                       'parents': None,
                       # identity of parent(s)
                       'age': 0.0,
                       # age of the organism
                       'gender': None,
                       # gender of organism
                       'lifespan': 100.0,
                       # maximum lifespan
                       'fitness': 100.0,
                       # % of fitness
                       'blood': None,
                       # interpreted chromosome
                       'identity': None,
                       # name of the organism
                       'deme': None,
                       # sub-population or race
                       'location': None,
                       # location of the organism
                       'death': None}
        if genome == 'dummy':
            self.genome = [Chromosome([0], [0])]
        else:
            self.genome = genome
        self.mutation_type = mutation_type
        self.additional_mutation_rate = \
            additional_mutation_rate
```

```
        self.status['gender'] = gender

    def generate_name(self):
        name = ''.join([random.choice(('1', '2', '3',
                '4', '5', '6', '7', '8', '9', 'A', 'B',
                'C', 'D', 'E', 'F', 'G', 'H', 'I', 'J',
                'K', 'L', 'M', 'N', 'O', 'P', 'Q', 'R',
                'S', 'T', 'U', 'V', 'W', 'X', 'Y', 'Z',
                'a', 'b', 'c', 'd', 'e', 'f', 'g', 'h',
                'i', 'j', 'k', 'l', 'm', 'n', 'o', 'p',
                'q', 'r', 's', 't', 'u', 'v', 'w', 'x',
                'y', 'z', '=', '#', '$', '%', '&', '@',
                '<', '>', '?')) for x in range(32)])
        self.status['identity'] = name

    def fitness(self):
    """Function to calculate the fitness of the current
    organism. B{This function MUST be over-ridden by the
    inherited class or substituted as fitness function is
    highly dependent on utility.} The only requirement is
    that a fitness score must be returned.

    Here, the sample implementation calculates fitness as
    proportion of the genome with '1's.

    @return: fitness score or fitness list

    @since: version 0.4"""
        one_count = sum([sum(chromosome.sequence)
                         for chromosome in self.genome])
        length_of_genome = sum([len(chromosome.sequence)
                          for chromosome in self.genome])
        return float(one_count) / float(length_of_genome)

    def mutation_scheme(self, type=None, rate=None):
    """Function to trigger mutation events in each
    chromosome. B{This function may be over-ridden by the
    inherited class or substituted to cater for specific
    mutation schemes but not an absolute requirement to
    do so.}

    Both type and rate must be defined at the same time,
    otherwise the initiated mutation_type and
    additional_mutation_rate will be used.

    @param type: type of mutation. Accepts 'point' (point
    mutation), 'insert' (insert a base), 'delete' (delete
    a base), 'invert' (invert a stretch of the
```

```
chromosome), 'duplicate' (duplicate a stretch of the
chromosome), 'translocate' (translocate a stretch of
chromosome to another random position). Default =
None.
@param rate: probability of mutation per base above
background mutation rate. Default = None. No mutation
event will ever happen if  (rate +
background_mutation) is less than zero.

@since: version 0.4"""
    for chromosome in self.genome:
        if not type and not rate:
            chromosome.rmutate(self.mutation_type,
                     self.additional_mutation_rate)
        if type and rate:
            chromosome.rmutate(type, rate)

def setStatus(self, variable, value):
"""Sets new status or change status of the organism.
However, the following status change will result in
death of the organism
    1. 'alive' to False
    2. 'vitality' to or below zero
    3. 'age' to or above lifespan

@param variable: name of status to change
@param value: new value of the status

@since: version 0.4"""
    if variable == 'alive' and value == True:
        self.status['alive'] = value
    if variable == 'alive' and value == False:
        self.status['alive'] = value
        self.status['death'] = 'death03'
    if variable == 'vitality':
        if float(value) > 100.1:
            self.status['vitality'] = 100.0
        if float(value) > 0.0 and \
           float(value) < 100.1:
            self.status['vitality'] = float(value)
        if float(value) == 0 or float(value) < 0:
            self.status['vitality'] = 0
            self.status['alive'] = False
            self.status['death'] = 'death01'
    elif variable == 'age':
        if float(value) < self.status['lifespan']:
            self.status['age'] = float(value)
        else:
```

```
                self.status['age'] = \
                    self.status['lifespan']
                self.status['alive'] = False
                self.status['death'] = 'death02'
        else:
            self.status[variable] = value

    def getStatus(self, variable):
    """Returns a status variable of the current organism

    @param variable: name of the status variable
    @return: status or a KeyError if status is not found

    @since: version 0.4"""
        try:
            return self.status[variable]
        except:
            raise KeyError('%s not found in organism \
                                status' % str(variable))

    def __str__(self):
    """Returns the genome of the organism"""
        return str([chromosome.sequence
                    for chromosome in self.genome])

    def clone(self):
    """Clones (deep copy) the organism.

    @return: a copy of current organism.

    @since: version 0.4"""
        org = deepcopy(self)
        return org

class Population(object):
"""Representation of a population as a list of organisms.
The entire population is in memory.

Methods to be over-ridden in the inherited class or
substituted are
    - prepopulation_control
    - mating
    - postpopulation_control
    - generation_events
    - report
```

```
@see: Lim, JZR, Aw, ZQ, Goh, DJW, How, JA, Low, SXZ, Loo,
BZL, Ling, MHT. 2010. A genetic algorithm framework
grounded in biology. The Python Papers Source Codes 2: 6.

@since: version 0.4"""

    def __init__(self, goal,
                 maxgenerations='infinite', agents=[]):
    """Establishes a population of organisms.

    @param goal: the goal to be reached by the
    population.
    @type goal: the return type of Organism.fitness()
    @param maxgenerations: maximum number of generations
    to evolve. Default = 'infinite'.
    @param agents: organisms making up the initial
    population.
    @type agents: list of Organism objects

    @since: version 0.4"""
        self.agents = agents
        self.goal = goal
        self.maxgenerations = maxgenerations
        self.generation = 0

    def prepopulation_control(self):
    """Function to trigger population control events
    before mating event in each generation (For example,
    to simulate pre-puberty death). B{This function may
    be over-ridden by the inherited class or substituted
    to cater for specific events.} Although this is not
    an absolute requirement, it is extremely encouraged
    to prevent exhaustion of memory space. Without
    population control, it will seem like a reproducing
    immortal population.

    Here, the sample implementation eliminates the bottom
    half of the population based on fitness score unless
    the number of organism is less than 20. This is to
    prevent extinction. However, if the number of
    organism is more than 2000 (more than 100x initial
    population size, a random selection of 2000 will be
    used for the next generation.

    @since: version 0.4"""
        size = len(self.agents)
        sfitness = [self.agents[x].fitness()
                    for x in range(size)]
```

```
        threshold = sum(sfitness) / len(sfitness)
        temp = [self.agents[x]
                for x in range(size)
                    if sfitness[x] > threshold]
        if len(temp) > 2001:
            size = len(temp)
            self.agents = [temp[random.randint(0,
                                    size - 1)]
                           for x in range(2000)]
        if len(temp) > 21:
            self.agents = temp

    def mating(self):
    """Function to trigger mating events in each
    generation. B{This function may be over-ridden by the
    inherited class or substituted to cater for specific
    mating schemes but not an absolute requirement to do
    so.} Mating schemes should include
        - selection of mating partners using
        Organism.fitness() function and/or other status
        - processes and actions of mating

    Here, the sample implementation randomly selects any
    2 organisms from the culled list and generate a new
    genome for the progeny organism by single crossover.

    @since: version 0.4"""
        size = len(self.agents)
        temp = []
        for x in range(size):
            organism1 = self.agents[random.randint(0,
                                        size - 1)]
            organism2 = self.agents[random.randint(0,
                                        size - 1)]
            crossover_pt = random.randint(0,
                 len(organism1.genome[0].sequence))
            (g1, g2) = crossover(organism1.genome[0],
                                 organism2.genome[0],
                                 crossover_pt)
            temp = temp + [Organism([g1])]
        self.add_organism(temp)

    def postpopulation_control(self):
    """Function to trigger population control events
    after mating event in each generation(For example, to
    simulate old-age death). B{This function may be over-
    ridden by the inherited class or substituted to cater
    for specific events.} Although this is not an
```

```
absolute requirement, it is extremely encouraged to
prevent exhaustion of memory space. Without
population control, it will seem like a reproducing
immortal population.

@since: version 0.4"""
    pass

def generation_events(self):
"""Function to trigger other defined events in each
generation. B{This function may be over-ridden by the
inherited class or substituted to cater for specific
events but not an absolute requirement to do so.}
Events and controls may include
    - processes simulating disaster or other
    catastrophic events
    - changes in mutations

@since: version 0.4"""
    pass

def report(self):
"""Function to report the status of each generation.
B{This function may be over-ridden by the inherited
class or substituted to cater for specific reporting
schemes but not an absolute requirement to do so.} At
the very least, this function should report whether
the goal is reached.

@return: dictionary of status describing the current
generation

@since: version 0.4"""
    sfitness = [self.agents[x].fitness()
                for x in range(len(self.agents))]
    afitness = sum(sfitness) / \
               float(len(self.agents))
    return {'generation': self.generation,
            'average fitness': afitness,
            '% to goal': float(afitness - self.goal)\
                          /  self.goal * 100}

def generation_step(self):
"""Function to simulate events for one generation.
These includes
    - mating (according to the mating scheme or
    function)
    - mutating each organism in the population
```

```
        - population size control
        - other events defined under generation_events
        function
        - increment of generation count
        - reporting the population status

    @return: information returned from report function.

    @since: version 0.4"""
        if self.generation > 0:
            self.prepopulation_control()
        self.mating()
        self.postpopulation_control()
        for organism in self.agents:
            organism.mutation_scheme()
        self.generation_events()
        self.generation = self.generation + 1
        return self.report()

    def add_organism(self, organism):
        """Add a new organism(s) to the population.

        @param organism: list of new Organism object(s)

        @since: version 0.4"""
        self.agents = self.agents + organism

    def freeze(self, prefix='pop', proportion=0.01):
    """Preserves part or the entire population. If the
    population size or the preserved proportion is below
    100, the entire population will be preserved. The
    preserved sample will be written into a file with
    name in the following format - <prefix><generation
    count>_<sample size>.gap

    @param prefix: prefix of file name. Default = 'pop'.
    @param proportion: proportion of population to be
    preserved. Default = 0.01, preserves 1% of the
    population.

    @since: version 0.4"""
        try:
            import cPickle as pickle
        except ImportError:
            import pickle
        if proportion > 1.0: proportion = 1.0
        if len(self.agents) < 101 or \
           len(self.agents) * proportion < 101:
```

```
            sample = self.agents
        else:
            size = len(self.agents)
            sample = [self.agents[random.randint(0,
                                  size - 1)]
                      for x in range(int(len(self.agents)
                          * proportion))]
        name = ''.join([prefix, str(self.generation),
                        '_', str(len(sample)), '.gap'])
        f = open(name, 'wb')
        pickle.dump(sample, f)
        f.close()

    def revive(self, filename, type='replace'):
    """Revives a frozen population.

    @param filename: file name of frozen population
    (generated by Population.freeze() function)
    @param type: type of revival. Allows 'replace'
    (replace the current population with the revived
    population) or 'add' (add the revived population to
    the current population). Default = 'replace'.

    @since: version 0.4"""
        try:
            import cPickle as pickle
        except ImportError:
            import pickle
        if type == 'replace':
            self.agents = \
                pickle.load(open(filename, 'rb'))
        if type == 'add':
            self.agents = self.agents + \
                 pickle.load(open(filename, 'rb'))

def crossover(chromosome1, chromosome2, position):
"""Cross-over operator - swaps the data on the 2 given
chromosomes after a given position.

@param chromosome1: Chromosome object
@param chromosome2: Chromosome object
@param position: base position of the swap over
@type position: integer
@return: (resulting chromosome1, resulting chromosome2)

New chromosomes inherit their parent's crossed sequences,
bases and background mutation rate.
```

```
@since: version 0.4"""
    seq1 = chromosome1.sequence
    seq2 = chromosome2.sequence
    position = int(position)
    if len(seq1) > position and len(seq2) > position:
        new1 = Chromosome(seq1[:position] + \

                               seq2[position:],
                               chromosome1.base,
                            chromosome1.background_mutation)
        new2 = Chromosome(seq2[:position] + \
                               seq1[position:],
                               chromosome2.base,
                            chromosome2.background_mutation)
        return (new1, new2)
    elif len(seq1) > position:
        new1 = Chromosome(seq1[:position],
                               chromosome1.base,
                            chromosome1.background_mutation)
        new2 = Chromosome(chromosome2.sequence + \
                               seq1[position:],
                               chromosome2.base,
                            chromosome2.background_mutation)
        return (new1, new2)
    elif len(seq2) > position:
        new1 = Chromosome(chromosome1.sequence + \
                               seq2[position:],
                               chromosome1.base,
                            chromosome1.background_mutation)
        new2 = Chromosome(seq2[:position],
                               chromosome2.base,
                            chromosome2.background_mutation)
        return (new1, new2)
    else:
        return (chromosome1, chromosome2)

population_data = \
{
    'chromosome_bases' : [1, 2, 3, 4],
    'chromosome_length' : 200,
    'chromosome_type' : 'defined',
    'initial_chromosome' : [1] * 200,
    'background_mutation' : 0.0001,
    'genome_size' : 1,
    'population_size' : 200,
    'fitness_function' : 'default',
    'mutation_scheme' : 'default',
    'additional_mutation' : 0.01,
```

```
    'mutation_type' : 'point',
    'goal' : 4,
    'maximum_generations' : 'infinite',
    'prepopulation_control' : 'default',
    'mating' : 'default',
    'postpopulation_control' : 'default',
    'generation_events' : 'default',
    'population_report' : 'default'
}

def population_constructor(data=population_data):
"""Function to construct a population based on a
dictionary of population data.

Population data contains the following keys:
    - 'chromosome_bases' = List of allowable bases.
          Default = [1, 2, 3, 4].
    - 'chromosome_length' = Length of a chromosome.
    Default = 200.
    - 'chromosome_type' = Type of chromosome. Default =
    'defined'.
    - 'initial_chromosome' = Initial chromosome. Default
    = [1] * 200.
    - 'background_mutation' = Background mutation rate.
    Default = 0.0001 (0.01%).
    - 'genome_size' = Number of chromosomes per organism.
    Default = 1.
    - 'population_size' = Size of initial population
    (number of organisms). Default = 200.
    - 'fitness_function' = Fitness evaluation function.
    Accepts 'default' or a function. Please refer to
    Organism. Default = 'default'.
    - 'mutation_scheme' = Function simulating the
    mutation scheme of an organism. Accepts 'default' or
    a function. Please refer to Organism. Default =
    'default'.
    - 'additional_mutation' = Mutation rate on top of
    background mutation rate. Default = 0.01 (1%).
    - 'mutation_type' = Type of default mutation. Default
    = 'point'.
    - 'goal' = Goal of the population, as evaluated by
    fitness function. Default = 4.
    - 'maximum_generations' = Number of generations to
    simulate. Accepts an integer or 'infinite'. Default =
    'infinite'.
    - 'prepopulation_control' = Function simulating pre-
    mating population control. Please refer to
    Population. Default = 'default'.
```

```
    - 'mating' = Function simulating mating procedure
    (mate selection and act of mating control. Please
    refer to Population. Default = 'default'.
    - 'postpopulation_control' = Function simulating
    post-mating population control. Please refer to
    Population. Default = 'default'.
    - 'generation_events' = Function simulating other
    possible (usually rare or random) events in the
    generation. Please refer to Population. Default =
    'default'.
    - 'population_report' = Function to generate the
    status report of the generation. Please refer to
    Population. Default = 'default'.

@param data: population data
@type data: dictionary

@return: Population object
@see: Lim, JZR, Aw, ZQ, Goh, DJW, How, JA, Low, SXZ, Loo,
BZL, Ling, MHT. 2010. A genetic algorithm framework
grounded in biology. The Python Papers Source Codes 2: 6.

@since: version 0.4"""
    chr = Chromosome(data['initial_chromosome'],
                     data['chromosome_bases'],
                     data['background_mutation'])
    org = Organism([chr]*data['genome_size'],
                   data['mutation_type'],
                   data['additional_mutation'])
    org_set = [org.clone()
               for x in range(data['population_size'])]
    pop = Population(data['goal'],
                     int(data['maximum_generations']),
                     org_set)
    return pop

def population_simulate(population,
                        printfreq=100,
                        freezefreq='never',
                        freezefile='pop',
                        freezeproportion=0.01,
                        resultfile='result.txt'):
"""Function to simulate the population - start the GA.

@param population: Population to run.
@type population: Population object
@param printfreq: Reporting intervals on screen. Default =
100.
```

```
@param freezefreq: Generation intervals to freeze
population into a file. See Population.freeze method.
Accepts an integer or 'never'. Default = 'never'.
@param freezefile: Prefix of file name of frozen
population. See Population.freeze method. Default = 'pop'.
@param freezeproportion: Proportion of population to be
preserved. See Population.freeze method. Default = 0.01,
preserves 1% of the population.
@param resultfile: Name of file to print out results of
each generation. Format of output is dependent on
reporting method of the population (Population.report). If
'None', it will print out the results into a file. Default
= 'result.txt'.
@see: Lim, JZR, Aw, ZQ, Goh, DJW, How, JA, Low, SXZ, Loo,
BZL, Ling, MHT. 2010. A genetic algorithm framework
grounded in biology. The Python Papers Source Codes 2: 6.

@since: version 0.4
"""
    if freezefreq == 'never': freezefreq = int(12e14)
    if resultfile != 'None':
        result = open(resultfile, 'w')
    report = population.generation_step()
    reportitems = ['|'.join([str(key), str(report[key])])
                   for key in list(report.keys())]
    result.writelines('|'.join(reportitems))
    result.writelines(os.linesep)
    while population.generation < \
        population.maxgenerations:
        if resultfile != 'None':
            report = population.generation_step()
            reportitems = ['|'.join([str(key),
                                     str(report[key])])
                           for key in list(report.keys())]
            result.writelines('|'.join(reportitems))
        if population.generation % int(printfreq) == 0:
            print('|'.join(reportitems))
        result.writelines(os.linesep)
        if population.generation % int(freezefreq) == 0:
            population.freeze(freezefile,
                              freezeproportion)
    population.freeze(freezefile, 1.0)
    result.close()
```

Examples of Simulation File from Castillo and Ling (2014a)

File Name: run_examples_without_installation.py

```
'''This file contains inserts the path for DOSE into
system path to allow for any examples importing this file
to execute without needing DOSE to be installed into into
Python site-packages.'''
import sys
import os

cwd = os.getcwd().split(os.sep)
#cwd[-1] = 'dose'
cwd = os.sep.join(cwd[:-1])

sys.path.append(cwd)

import dose
```

File Name: 09_no_migration_isolated_mating.py

```
'''Example 09a: Examining the effects of no migration on
genetic distance differences from an initially identical
population (development of sub-populations or demes which
may lead to speciation)

This example is identical to Example 03, except background
mutation rate is changed from 10% in Example 03 to 0.1% in
this example.

In this simulation,
    - 1 populations of 1250 organisms
    - each organism will have 1 chromosome of only 2
    bases (1 and 0)
    - Evenly deployed across 25 eco-cells (50 organism
    per eco-cell)
    - 0.1% background point mutation on chromosome of 50
    bases
    - no organism movement throughout the simulation
    - no Ragaraja interpretation of genome
    - 1000 generations to be simulated'''
# needed to run this example without prior
```

```
# installation of DOSE into Python site-packages
try:
    import run_examples_without_installation
except ImportError: pass

# Example codes starts from here
import dose, random

parameters = \
{"simulation_name": "09_no_migration_isolated_mating",
 "population_names": ['pop_01'],
 "population_locations": [[(x,y,z)
                          for x in range(5)
                              for y in range(5)
                                  for z in range(1)]],
 "deployment_code": 3,
 "chromosome_bases": ['0','1'],
 "background_mutation": 0.001,
 "additional_mutation": 0,
 "mutation_type": 'point',
 "chromosome_size": 5000,
 "genome_size": 1,
 "max_tape_length": 50,
 "clean_cell": True,
 "interpret_chromosome": False,
 "max_codon": 2000,
 "population_size": 1250,
 "eco_cell_capacity": 50,
 "world_x": 5,
 "world_y": 5,
 "world_z": 1,
 "goal": 0,
 "maximum_generations": 1000,
 "fossilized_ratio": 0.01,
 "fossilized_frequency": 100,
 "print_frequency": 10,
 "ragaraja_version": 0,
 "ragaraja_instructions": ['000', '001', '010',
                           '011', '100', '101'],
 "eco_buried_frequency": 1250,
 "database_file": "sim09_no_migration.db",
 "database_logging_frequency": 1}

class simulation_functions(dose.dose_functions):

    def organism_movement(self, Populations,
        pop_name, World): pass
    def organism_location(self, Populations,
```

```
        pop_name, World): pass
    def ecoregulate(self, World): pass
    def update_ecology(self, World, x, y, z): pass
    def update_local(self, World, x, y, z): pass
    def report(self, World): pass
    def fitness(self, Populations, pop_name): pass

    def mutation_scheme(self, organism):
        organism.genome[0].rmutate(
           parameters["mutation_type"],
           parameters["additional_mutation"])

    def prepopulation_control(self, Populations,
        pop_name): pass

    def mating(self, Populations, pop_name):
        for location in parameters
            ["population_locations"][0]:
            group = dose.filter_location(location,
                    Populations[pop_name].agents)
            for x in range(len(group)//2):
                parents = []
                for i in range(2):
                    parents.append(random.choice(
                        Populations[pop_name].agents))
                    while parents[i] not in group:
                        parents[i] = \
                            random.choice(
                            Populations[pop_name].agents)
                    Populations[pop_name].agents.

                    remove(parents[i])
                crossover_pt = \
                    random.randint(0,
                    len(parents[0].genome[0].sequence))
                (new_chromo1, new_chromo2) = \
                    dose.genetic.crossover(
                    parents[0].genome[0],
                    parents[1].genome[0], crossover_pt)
                children = \
                  [dose.genetic.Organism([new_chromo1],
                      parameters["mutation_type"],
                      parameters["additional_mutation"]),
                   dose.genetic.Organism([new_chromo2],
                      parameters["mutation_type"],
                      parameters["additional_mutation"])]
                for child in children:
                    child.status['parents'] = \
```

```
                    [parents[0].status['identity'],
                     parents[1].status['identity']]
                child.status['location'] = location
                child.generate_name()
                child.status['deme'] = pop_name
                Populations[pop_name].
                    agents.append(child)

    def postpopulation_control(self, Populations,
        pop_name): pass
    def generation_events(self, Populations,
        pop_name): pass

    def population_report(self, Populations, pop_name):
        report_list = []
        for organism in Populations[pop_name].agents:
            identity = str(organism.status['identity'])
            report_list.append(identity)
        return '\n'.join(report_list)

    def database_report(self, con, cur, start_time,
                        Populations, World,
                        generation_count):
        try:
            dose.database_report_populations(con, cur,
               start_time, Populations, generation_count)
        except: pass
        try:
            dose.database_report_world(con, cur,
               start_time, World, generation_count)
        except: pass

    def deployment_scheme(self, Populations,
        pop_name, World): pass

dose.simulate(parameters, simulation_functions)
```

File Name: 09_adjacent_migration_isolated_mating.py

```
'''Example 09b: Extending from Example 09a (as baseline)
to examine the effects of short migration (adjacent cell
migration)  on genetic distance differences from an
initially identical population (development of sub-
populations or demes which may lead to speciation)
```

```
This example is identical to Example 04, except background
mutation rate is changed from 10% in Example 03 to 0.1% in
this example.

In this simulation,
    - 1 populations of 1250 organisms
    - each organism will have 1 chromosome of only 2
    bases (1 and 0)
    - Evenly deployed across 25 eco-cells (50 organism
    per eco-cell)
    - 0.1% background point mutation on chromosome of 50
    bases
    - 10% organism movement per eco-cell per generation
    throughout the simulation
    - the same organism may move twice due to sequential
    evaluation of the eco-cells but the probability of
    such event will be 10% x 10% = 1%; similarly, 3 or
    more movement by the same organism may happen with
    reducing probabilities
    - no Ragaraja interpretation of genome
    - 1000 generations to be simulated'''
# needed to run this example without prior
# installation of DOSE into Python site-packages
try:
    import run_examples_without_installation
except ImportError: pass
# Example codes starts from here
import dose, random

parameters = \
{"simulation_name": \
    "09_adjacent_migration_isolated_mating",
 "population_names": ['pop_01'],
 "population_locations": [[(x,y,z)
                           for x in range(5)
                               for y in range(5)
                                   for z in range(1)]],
 "deployment_code": 3,
 "chromosome_bases": ['0','1'],
 "background_mutation": 0.001,
 "additional_mutation": 0,
 "mutation_type": 'point',
 "chromosome_size": 5000,
 "genome_size": 1,
 "max_tape_length": 50,
 "clean_cell": True,
 "interpret_chromosome": False,
 "max_codon": 2000,
```

```
"population_size": 1250,
"eco_cell_capacity": 50,
"world_x": 5,
"world_y": 5,
"world_z": 1,
"goal": 0,
"maximum_generations": 1000,
"fossilized_ratio": 0.01,
"fossilized_frequency": 100,
"print_frequency": 10,
"ragaraja_version": 0,
"ragaraja_instructions": ['000', '001', '010',
                          '011', '100', '101'],
"eco_buried_frequency": 1250,
"database_file": "sim09_adjacent.db",
"database_logging_frequency": 1}

class simulation_functions(dose.dose_functions):

    def organism_movement(self, Populations,
        pop_name, World):
        for location in parameters
            ["population_locations"][0]:
            group = dose.filter_location(location,
                Populations[pop_name].agents)
            adj_cells = dose.simulation_calls.
                adjacent_cells(parameters, location)
            for i in range(int(round((len(group) \
                * 0.1)))):
                (x,y,z) = dose.simulation_calls. \
                          coordinates(location)
                World.ecosystem[x][y][z]['organisms'] =\
                World.ecosystem[x][y][z]['organisms'] - 1
                immigrant = random.choice(
                    Populations[pop_name].agents)
                while immigrant not in group:
                    immigrant = random.choice(
                       Populations[pop_name].agents)
                new_location = random.choice(adj_cells)
                immigrant.status['location'] = \
                    new_location
                (x,y,z) = dose.simulation_calls. \
                          coordinates(new_location)
                World.ecosystem[x][y][z]['organisms'] =\
                World.ecosystem[x][y][z]['organisms'] + 1

    def organism_location(self, Populations,
        pop_name, World): pass
```

```
def ecoregulate(self, World): pass
def update_ecology(self, World, x, y, z): pass
def update_local(self, World, x, y, z): pass
def report(self, World): pass
def fitness(self, Populations, pop_name): pass

def mutation_scheme(self, organism):
    organism.genome[0].rmutate(
        parameters["mutation_type"],
        parameters["additional_mutation"])

def prepopulation_control(self, Populations,
    pop_name): pass

def mating(self, Populations, pop_name):
    for location in parameters
        ["population_locations"][0]:
        group = dose.filter_location(location,
                Populations[pop_name].agents)
        for x in range(len(group)//2):
            parents = []
            for i in range(2):
                parents.append(
                    random.choice(
                    Populations[pop_name].agents))
                while parents[i] not in group:
                    parents[i] = random.choice(
                        Populations[pop_name].agents)
                Populations[pop_name].agents.
                    remove(parents[i])
            crossover_pt = random.randint(0,
                len(parents[0].genome[0].sequence))
            (new_chromo1, new_chromo2) = \
                dose.genetic.crossover(
                parents[0].genome[0],
                parents[1].genome[0], crossover_pt)
            children = [dose.genetic.Organism(
                        [new_chromo1],
                        parameters["mutation_type"],
                  parameters["additional_mutation"]),
                       dose.genetic.Organism(
                        [new_chromo2],
                        parameters["mutation_type"],
                  parameters["additional_mutation"])]
            for child in children:
                child.status['parents'] = \
                    [parents[0].status['identity'],
                     parents[1].status['identity']]
```

```
                    child.status['location'] = location
                    child.generate_name()
                    child.status['deme'] = pop_name
                    Populations[pop_name].
                        agents.append(child)

    def postpopulation_control(self, Populations,
        pop_name): pass
    def generation_events(self, Populations,
        pop_name): pass

    def population_report(self, Populations, pop_name):
        report_list = []
        for organism in Populations[pop_name].agents:
            identity = str(organism.status['identity'])
            report_list.append(identity)
        return '\n'.join(report_list)

    def database_report(self, con, cur, start_time,
                        Populations, World,
                        generation_count):
        try:
            dose.database_report_populations(con, cur,
              start_time, Populations, generation_count)
        except: pass
        try:
            dose.database_report_world(con, cur,
              start_time, World, generation_count)
        except: pass

    def deployment_scheme(self, Populations,
        pop_name, World): pass

dose.simulate(parameters, simulation_functions)
```

File Name: 09_long_migration_isolated_mating.py

```
'''Example 09c: Extending from Example 09a (as baseline)
and Example 09b (short distance migration) to examine the
effects of long distance migration (across one or more
eco-cells) on genetic distance differences from an
initially identical population (development of sub-
populations or demes which may lead to speciation)
```

```
This example is identical to Example 05, except background
mutation rate is changed from 10% in Example 03 to 0.1% in
this example.

In this simulation,
    - 1 populations of 1250 organisms
    - each organism will have 1 chromosome of only 2
    bases (1 and 0)
    - Evenly deployed across 25 eco-cells (50 organism
    per eco-cell)
    - 20% background point mutation on chromosome of 50
    bases
    - 10% organism movement per eco-cell per generation
    throughout the simulation
    - the same organism may move twice due to sequential
    evaluation of the eco-cells but the probability of
    such event will be 10% x 10% = 1%; similarly, 3 or
    more movement by the same organism may happen with
    reducing probabilities
    - the destination eco-cell for movement is random;
    thus, from 25 possible eco-cells, there is a
    probability of 4% chance of relocating back to the
    same (original) eco-cell, 16% chance of relocating to
    one of the 4-neighbour eco-cells and 32% chance of
    relocating to one of the 8-neighbour eco-cells
    - no Ragaraja interpretation of genome
    - 1000 generations to be simulated'''
# needed to run this example without prior
# installation of DOSE into Python site-packages
try:
    import run_examples_without_installation
except ImportError: pass

# Example codes starts from here
import dose, random

parameters = \
{"simulation_name": "09_long_migration_isolated_mating",
 "population_names": ['pop_01'],
 "population_locations": [[(x,y,z)
                           for x in range(5)
                               for y in range(5)
                                   for z in range(1)]],
 "deployment_code": 3,
 "chromosome_bases": ['0','1'],
 "background_mutation": 0.001,
 "additional_mutation": 0,
 "mutation_type": 'point',
```

```
 "chromosome_size": 5000,
 "genome_size": 1,
 "max_tape_length": 50,
 "clean_cell": True,
 "interpret_chromosome": False,
 "max_codon": 2000,
 "population_size": 1250,
 "eco_cell_capacity": 50,
 "world_x": 5,
 "world_y": 5,
 "world_z": 1,
 "goal": 0,
 "maximum_generations": 1000,
 "fossilized_ratio": 0.01,
 "fossilized_frequency": 100,
 "print_frequency": 10,
 "ragaraja_version": 0,
 "ragaraja_instructions": ['000', '001', '010',
                           '011', '100', '101'],
 "eco_buried_frequency": 1250,
 "database_file": "sim09_long_migration.db",
 "database_logging_frequency": 1}

class simulation_functions(dose.dose_functions):

    def organism_movement(self, Populations,
        pop_name, World): pass

    def organism_location(self, Populations,
        pop_name, World):
        for location in parameters
            ["population_locations"][0]:
            group = dose.filter_location(location,
                    Populations[pop_name].agents)
            for i in range(int(round((len(group) * \
                0.1)))):
                (x,y,z) = dose.simulation_calls.
                          coordinates(location)
                World.ecosystem[x][y][z]['organisms'] =\
                World.ecosystem[x][y][z]['organisms'] - 1
                immigrant = random.choice(
                    Populations[pop_name].agents)
                while immigrant not in group:
                    immigrant = random.choice(
                        Populations[pop_name].agents)
                new_location = \
                    random.choice(parameters
                    ["population_locations"][0])
```

```
            while new_location == location:
                new_location = \
                    random.choice(parameters
                    ["population_locations"][0])
            immigrant.status['location'] = \
                new_location
            (x,y,z) = dose.simulation_calls.
                      coordinates(new_location)
            World.ecosystem[x][y][z]['organisms'] =\
            World.ecosystem[x][y][z]['organisms'] + 1

    def ecoregulate(self, World): pass
    def update_ecology(self, World, x, y, z): pass
    def update_local(self, World, x, y, z): pass
    def report(self, World): pass
    def fitness(self, Populations, pop_name): pass

    def mutation_scheme(self, organism):
        organism.genome[0].rmutate
           (parameters["mutation_type"],
            parameters["additional_mutation"])

    def prepopulation_control(self, Populations,
        pop_name): pass

    def mating(self, Populations, pop_name):
        for location in parameters
           ["population_locations"][0]:
            group = dose.filter_location(location,
                    Populations[pop_name].agents)
            for x in range(len(group)//2):
                parents = []
                for i in range(2):
                    parents.append(random.choice(
                    Populations[pop_name].agents))
                    while parents[i] not in group:
                        parents[i] = \
                            random.choice(
                            Populations[pop_name].agents)
                    Populations[pop_name].agents.
                        remove(parents[i])
                crossover_pt = random.randint(0,
                    len(parents[0].genome[0].sequence))
                (new_chromo1, new_chromo2) = \
                    dose.genetic.crossover(
                        parents[0].genome[0],
                        parents[1].genome[0],
                        crossover_pt)
```

```
            children = [dose.genetic.Organism(
                         [new_chromo1],
                         parameters["mutation_type"],
                 parameters["additional_mutation"]),
                        dose.genetic.Organism(
                         [new_chromo2],
                         parameters["mutation_type"],
                 parameters["additional_mutation"])]
            for child in children:
                child.status['parents'] = \
                    [parents[0].status['identity'],
                    parents[1].status['identity']]
                child.status['location'] = location
                child.generate_name()
                child.status['deme'] = pop_name
                Populations[pop_name].agents.
                    append(child)

    def postpopulation_control(self, Populations,
        pop_name): pass
    def generation_events(self, Populations,
        pop_name): pass

    def population_report(self, Populations, pop_name):
        report_list = []
        for organism in Populations[pop_name].agents:
            identity = str(organism.status['identity'])
            report_list.append(identity)
        return '\n'.join(report_list)

    def database_report(self, con, cur, start_time,
                        Populations, World,
                        generation_count):
        try:
            dose.database_report_populations(con, cur,
               start_time, Populations, generation_count)
        except: pass
        try:
            dose.database_report_world(con, cur,
               start_time, World, generation_count)
        except: pass

    def deployment_scheme(self, Populations,
        pop_name, World): pass

dose.simulate(parameters, simulation_functions)
```

EXAMPLE FOR DEMONSTRATING REVIVAL OF SIMULATION FROM DATABASE

```
'''Reviving a simulation from simulation logging database
and run the simulation for another 200 generations.

Example 09 ran the simulation for 1000 generations and
logged the events into case_study_01.db file. This example
revived the logged simulation from case_study_01.db file
using the simulation start time as identifier (2013-10-19-
1382200534.1) and run the simulation from generation 1001
to 1200. The events of generation 1001 to 1200 are logged
in case_study_01.db
file.'''
try:
    import run_examples_without_installation
except ImportError: pass

import dose, genetic
import os, random

'''
Needs pre-existing 09_no_migration_isolated_mating.py.
Change simulation_time below to the corresponding starting
time of the said simulation.
'''
rev_parameters = \
{"database_source" : "case_study_01.db",
 "simulation_time": "2013-10-19-1382200534.1",
 "population_locations": [[(x,y,z)
                              for x in xrange(5)
                                  for y in xrange(5)
                                      for z in xrange(1)]],
 "rev_start" : [1000],
 "extend_gen" : 200,
 "simulation_name": "08_revive_simulation_03",
 "chromosome_bases": ['0','1'],
 "background_mutation": 0.1,
 "additional_mutation": 0,
 "mutation_type": 'point',
 "max_tape_length": 50,
 "clean_cell": True,
 "interpret_chromosome": True,
 "max_codon": 2000,
 "goal": 0,
 "eco_cell_capacity": 100,
```

```
 "fossilized_ratio": 0.01,
 "fossilized_frequency": 20,
 "print_frequency": 10,
 "ragaraja_version": 0,
 "ragaraja_instructions": ['000', '001', '010',
                            '011', '100', '101'],
 "eco_buried_frequency": 100,
 "database_file": "case_study_01.db",
 "database_logging_frequency": 1}

class simulation_functions(dose.dose_functions):

    def organism_movement(self, Populations,
        pop_name, World): pass
    def organism_location(self, Populations,
        pop_name, World): pass
    def ecoregulate(self, World): pass
    def update_ecology(self, World, x, y, z): pass
    def update_local(self, World, x, y, z): pass
    def report(self, World): pass
    def fitness(self, Populations, pop_name): pass

    def mutation_scheme(self, organism):
        organism.genome[0].rmutate(
            rev_parameters["mutation_type"],
            rev_parameters["additional_mutation"])

    def prepopulation_control(self, Populations,
        pop_name): pass
    def mating(self, Populations, pop_name):
        for location in rev_parameters
            ["population_locations"][0]:
            group = dose.filter_location(location,
                    Populations[pop_name].agents)
            for x in xrange(len(group)/2):
                parents = []
                for i in xrange(2):
                    parents.append(random.choice(
                        Populations[pop_name].agents))
                    while parents[i] not in group:
                        parents[i] = random.choice(
                            Populations[pop_name].agents)
                    Populations[pop_name].agents.
                        remove(parents[i])
                crossover_pt = random.randint(0,
                    len(parents[0].genome[0].sequence))
                (new_chromo1, new_chromo2) = \
                    genetic.crossover(
```

```
                    parents[0].genome[0],
                    parents[1].genome[0],
                    crossover_pt)
            children = [genetic.Organism(
                    [new_chromo1],
                    rev_parameters["mutation_type"],
             rev_parameters["additional_mutation"]),
                  genetic.Organism(
                    [new_chromo2],
                     rev_parameters["mutation_type"],
             rev_parameters["additional_mutation"])]
            for child in children:
                child.status['location'] = location
                child.generate_name()
                child.status['deme'] = pop_name
                Populations[pop_name].agents.
                    append(child)

    def postpopulation_control(self, Populations,
        pop_name): pass
    def generation_events(self, Populations,
        pop_name): pass

    def population_report(self, Populations, pop_name):
        report_list = []
        for organism in Populations[pop_name].agents:
            chromosome = \
                ''.join(organism.genome[0].sequence)
            location = str(organism.status['location'])
            report_list.append(chromosome + '  ' + \
                               location)
        return '\n'.join(report_list)

    def database_report(self, con, cur, start_time,
                        Populations, World,
                        generation_count):
        try:
            dose.database_report_populations(con, cur,
               start_time, Populations, generation_count)
        except: pass
        try:
            dose.database_report_world(con, cur,
               start_time, World, generation_count)
        except: pass

    def deployment_scheme(self, Populations,
        pop_name, World): pass
```

```
dose.revive_simulation(rev_parameters,
                       simulation_functions)
```

How to Write a Simulation

A simulation script is a Python code with 4 sections:

1) Import dose library by "import dose"
2) A dictionary of simulation parameters
3) A class containing simulation methods/functions
4) Run simulation by calling DOSE simulation function with the simulation parameters and simulation methods as parameters: dose.simulate(parameters, simulation_functions)

Since (1) and (4) are standard, only (2) and (3) are elaborated here.

A dictionary of simulation parameters provides for the set of parameters needed to construct the simulation and store the simulation results.

```
parameters = \
{"simulation_name": \
    "01_basic_functions_one_cell_deployment",
 "population_names": ['pop_01'],
 "population_locations": [[(0,0,0)]],
 "deployment_code": 1,
 "chromosome_bases": ['0','1'],
 "background_mutation": 0.1,
 "additional_mutation": 0,
 "mutation_type": 'point',
 "chromosome_size": 30,
 "genome_size": 1,
 "max_tape_length": 50,
 "clean_cell": True,
 "interpret_chromosome": True,
 "max_codon": 2000,
 "population_size": 100,
 "eco_cell_capacity": 100,
 "world_x": 5,
 "world_y": 5,
```

```
 "world_z": 5,
 "goal": 0,
 "maximum_generations": 100,
 "fossilized_ratio": 0.01,
 "fossilized_frequency": 20,
 "print_frequency": 10,
 "ragaraja_version": 0,
 "ragaraja_instructions": ['000', '001', '010',
                           '011', '100', '101'],
 "eco_buried_frequency": 100,
 "database_file": "simulation.db",
 "database_logging_frequency": 1}
```

A class containing simulation methods/functions provides a set of simulation-specific functions.

```
class simulation_functions(dose.dose_functions):

    def organism_movement(self, Populations,
        pop_name, World): pass
    def organism_location(self, Populations,
        pop_name, World): pass
    def ecoregulate(self, World): pass
    def update_ecology(self, World, x, y, z): pass
    def update_local(self, World, x, y, z): pass
    def report(self, World): pass
    def fitness(self, Populations, pop_name): pass

    def mutation_scheme(self, organism):
        organism.genome[0].rmutate(
            parameters["mutation_type"],
            parameters["additional_mutation"])

    def prepopulation_control(self, Populations,
        pop_name): pass
    def mating(self, Populations, pop_name): pass
    def postpopulation_control(self, Populations,
        pop_name): pass
    def generation_events(self, Populations,
        pop_name): pass

    def population_report(self, Populations, pop_name):
        sequences = [''.join(org.genome[0].sequence)
            for org in Populations[pop_name].agents]
        identities = [org.status['identity']
            for org in Populations[pop_name].agents]
        locations = [str(org.status['location'])
            for org in Populations[pop_name].agents]
```

```
        demes = [org.status['deme']
                 for org in Populations[pop_name].agents]
        return '\n'.join(sequences)

    def database_report(self, con, cur, start_time,
                        Populations, World,
                        generation_count):
        try:
            dose.database_report_populations(con, cur,
              start_time, Populations, generation_count)
        except: pass
        try:
            dose.database_report_world(con, cur,
              start_time, World, generation_count)
        except: pass

    def deployment_scheme(self, Populations,
        pop_name, World): pass
```

References

Anderson, CJR, and Harmon, L. (2014). Ecological and Mutation-Order Speciation in Digital Organisms. *The American Naturalist 183*, 257–268.

Batut, B, Parsons, D, Fischer, S, Beslon, G, and Knibbe, C. (2013). In silico experimental evolution: a tool to test evolutionary scenarios. *BMC Bioinformatics 14*, S11.

Beckmann, BE, Knoester, DB, Connelly, BD, Waters, CM, and McKinley, PK. (2012). Evolution of Resistance to Quorum Quenching in Digital Organisms. *Artificial Life 18*, 291–310.

Bersini, H. (2009). How artificial life relates to theoretical biology. *Origins of Life: Self-Organization and/or Biological Evolution?* 61–78.

Buckling, A, Maclean, RC, Brockhurst, MA, and Colegrave, N. (2009). The Beagle in a bottle. *Nature 457*, 824–829.

Castillo, CF, and Ling, MH. (2014a). Digital Organism Simulation Environment (DOSE): a library for ecologically-based in silico experimental evolution. *Advances in Computer Science: An International Journal 3*, 44–50.

Castillo, CF, and Ling, MH. (2014b). Resistant Traits in Digital Organisms Do Not Revert Preselection Status despite Extended Deselection: Implications to Microbial Antibiotics Resistance. *BioMed Research International 2014*, 648389.

Castillo, CF, Chay, ZE, and Ling, MHT. (2015). Resistance Maintained in Digital Organisms despite Guanine/Cytosine-Based Fitness Cost and Extended De-Selection: Implications to Microbial Antibiotics Resistance. *MOJ Proteomics & Bioinformatics 2*, 00039.

Clarke, CA., Mani, G, and Wynne, G. (1985). Evolution in reverse: clean air and the peppered moth. *Biological Journal of the Linnean Society 26*, 189–199.

Gupta, A, LaBar, T, Miyagi, M, and Adami, C. (2016). Evolution of Genome Size in Asexual Digital Organisms. *Scientific Reports 6*, 25786.

Koh, YZ, and Ling, MH. (2013). On the Liveliness of Artificial Life. *iConcept Journal of Human-Level Intelligence 3*, 1.

LaBar, T, and Adami, C. (2016). Different Evolutionary Paths to Complexity for Small and Large Populations of Digital Organisms. *PLoS Comput. Biol. 12*, e1005066.

Langton, CG. (1986). Studying artificial life with cellular automata. *Physica D: Nonlinear Phenomena 22*, 120–149.

Lim, JZ, Aw, ZQ, Goh, DJ, How, JA, Low, SX, Loo, BZ, and Ling, MHT. (2010). A genetic algorithm framework grounded in biology. *The Python Papers Source Codes 2*, 6.

Ling, MHT. (2012a). An artificial life simulation library based on genetic algorithm, 3-character genetic code and biological hierarchy. *The Python Papers 7*, 5.

Ling, MHT. (2012b). Ragaraja 1.0: The Genome Interpreter of Digital Organism Simulation Environment (DOSE). *The Python Papers Source Codes 4*, 2.

Ling, MHT. (2014). Applications of Artificial Life and Digital Organisms in the Study of Genetic Evolution. *Advances in Computer Science: An International Journal 3*, 107–112.

Raymond, L, Plantegenest, M, and Vialatte, A. (2013). Migration and dispersal may drive to high genetic variation and significant genetic mixing: the case of two agriculturally important, continental hoverflies (*Episyrphus balteatus* and *Sphaerophoria scripta*). *Molecular Ecology 22*, 5329–5339.

Relethford, JH. (1988). Heterogeneity of long-distance migration in studies of genetic structure. *Annals of Human Biology 15*, 55–63.

Song, Y, Endepols, S, Klemann, N, Richter, D, Matuschka, FR, Shih, CH, Nachman, MW, and Kohn, MH. (2011). Adaptive introgression of anticoagulant rodent poison resistance by hybridization between old world mice. *Current Biology 21*, 1296–1301.

Stuart, YE, Campbell, T, Hohenlohe, P, Reynolds, RG, Revell, L, and Losos, J. (2014). Rapid evolution of a native species following invasion by a congener. *Science 346*, 463–466.

Date Submitted: May 13, 2017. *Date Accepted:* June 20, 2017.

In: Current STEM. Volume 1
Editor: Maurice HT Ling
ISBN: 978-1-53613-416-2

Chapter 2

SOCIO-CULTURAL AND BEHAVIORAL FACTORS INFLUENCING RISK OF HIV AMONG WOMEN IN GRENADA

Winta Ghidei[1,*] and Shelly Rodrigo[2,†]
[1]University of Alberta, School of Public Health, Edmonton, Canada
[2]St. George's University, School of Medicine, Department of Public Health and Preventive Medicine, St. George, Grenada

ABSTRACT

Identification of the major risk factors for HIV is key in preventing and controlling the spread of the virus. This cross-sectional study was the first to systematically investigate the social, cultural, economic and behavioral risk factors and challenges faced by Grenadian women making them susceptible to HIV/AIDS. A convenient sample of Grenadian women

* Corresponding Author Emails: ghidei@ualberta.ca, winta.ghidei@gmail.com.
† Corresponding Author Email: srodrigo@sgu.edu.

of reproductive age (17-45) were selected for the study. Data were collected using a structured questionnaire and were analyzed using the Statistical Package for Social Sciences (Version 22). Associations between sexual risk behaviors and several independent variables (socio-demographic factors, economically motivated relationships, influence of music genre, intimate partner violence) were examined, at the 5% significance level. Health risk behaviors were higher in women with lower educational levels and economic statuses. Music genres (dancehall, soca and reggae) and intimate partner violence also influenced sexual risk behaviors. The health risk behaviors leading to HIV are of concern for Grenadian women and require urgent attention. Addressing the issue at the social level would be most effective but could take time and needs more resources as well as political will at the national level. However, immediate attention can be taken at the individual and community levels.

Abbreviations

HIV: Human Immunodeficiency Virus
AIDS: Acquired Immune Deficiency Syndrome
STI: Sexually Transmitted Infection
WHO: World Health Organization
CDC: Center for Diseases Control

Background

In 2015, 36.7 million people were living with HIV (Human Immunodeficiency Virus) and 1.1 million died of AIDS (Acquired Immune Deficiency Syndrome) related complications worldwide [1]. Since the beginning of the HIV epidemic, 75 million people have been infected and 36 million died of the disease [1]; with 97% of those infected reported to reside in low and middle income countries [2].

The Caribbean region has the second highest HIV/AIDS rate in the world [3]. In this region, 230,000 people were reported to carry HIV in 2011, of whom 13,000 were infected that year, 10,000 died of the disease and half of the infected were women [4]. In Grenada, a country in the Eastern Caribbean (population number = 105,897), 303 people have been infected with HIV/AIDS since the beginning of the epidemic, and 220 died of AIDS [5]. Youth between the ages of 25 and 34 comprise 65% of those infected with HIV, and within this age group 34% were female [5].

Globally, women are disproportionally affected by HIV/AIDS compared to men due to behavioral and social factors. Those behavioral factors include having unprotected sex (not using condoms), having concurrent multiple partners, and engaging in transactional sex. Generally, social factors such as gender disparities, societal ideologies, and economy have an impact on these behavioral factors [6].

Gender disparities in most developing countries increase women's susceptibility to HIV due to their low social statuses within the communities [7]. Poverty, lack of education, unemployment, and the cultural and sexual norms of society play major roles in lowering women's social statuses [8, 9]. This lower social status leads to higher risk for rape, unprotected sex and sexual violence, making women more vulnerable to HIV/AIDS [7]. In addition, exposures to childhood sexual trauma and to intimate partner violence have impacted sexual health risk behaviors [8, 10].

In the Caribbean region, social norms encourage men to have multiple partners and women to start having sexual intercourse at a younger age [11]. Due to cultural pressures, girls under the age of 15 years old have sexual intercourse [7]. Additionally, gender inequalities in the region, and the low economic statuses within most of these countries expose women to transactional sex and reduce their chances of condom negotiation [11]. Age imbalances and gender-based physical and sexual violence also make women more vulnerable to HIV infection [11]. At the same time, individual or societal definitions of sexual pleasure, myths, religious views and doubts in the efficiency of condoms have a great effect on spread of the virus [12].

Although there is high prevalence of HIV in the Caribbean region, the socio-cultural risk factors, and the impact of social norms have not been investigated at the country levels; nor has the challenges faced by the women been investigated. This study examined the socio-cultural, economic and behavioral risk factors and challenges faced by Grenadian women that may impact their vulnerability to HIV/AIDS. Since women's health is a vital factor in a nation's development, identifying these risk factors and preventing them would not only benefit Grenadian women, but will ensure a positive and sustainable change to the nation in general. The information collected could facilitate the designing of appropriate and more tailored interventions to prevent HIV/AIDS in this population group.

METHODS

Study Setting

The study was conducted in the state of Grenada, located in the Eastern Caribbean region comprising three islands: Grenada, Carriacou and Petite Martinique. The main island, Grenada, consists of six parishes (Saint Patrick, Saint Mark, Saint John, Saint Andrew, Saint George, and Saint David), with the majority of the population residing in St George.

Study Samples

Participants were Grenadian women of reproductive age (17-45 years old) who were born in Grenada, and/or had lived on the island for the past 10 years. According to the Grenadian census data, the total population number of Grenadian females between the ages of 15 and 44 was 23,783 in 2011. Based in this number, a representative sample size of 380 was calculated at 80% power. However, a sample size of 400 was determined to be the minimum for surveying to account for 10% of the data being unusable due to lack of responses on questions relating to the independent and dependent variables.

Design and Procedure

In this cross-sectional study, convenience sampling was used to due to funding and time constrictions. However, participants were randomly chosen from colleges, work places, and public locations (markets, malls, etc.) in all six parishes and the island of Carriacou. Participants were recruited in person and through e-mails and flier distributions. The principal investigator administered a paper version of the questionnaire; and a link to the electronic version was provided for those who preferred this method.

The questionnaire was constructed using items utilized in similar studies [11, 13-17]. The survey items were previously validated; however, a pilot test was conducted to account for cultural competency.

Ten (10) participants from the different age groups were asked to participate in the pilot study. Five (5) of them were given the electronic version and the other five (5) were asked to complete the paper version of the questionnaire. After completing the questionnaire, participants were asked if they understood the meaning of each item, how they formulated their responses, if any emotions where evoked, and their general opinions related to its cultural competence [15]. The questionnaire was then edited accordingly.

Statistical Analysis

Data were analyzed using the Statistical Package for Social Sciences (SPSS), version 22.0. Chi square analysis was used to determine if there was a statistically significant difference in the levels of the demographic variables (Parish, Marital Status, Religion, Education, Occupation, and Living Status) for the following factors: music influence; family relationships and cultural beliefs; and health risk behaviors. Correlation was used to assess relationships between age of participants, age at first sexual intercourse and the appropriate age at which participants think a girl could start having intercourse. Binary logistic regression was also used to identify independent predictors of health risk behaviors. The level of significance was determined at an alpha level of 0.05.

Ethics

Ethical clearance and permission to proceed was obtained from the Internal Review Board (IRB) at St George's University School of Medicine, St George, Grenada. Informed consent was obtained from all participants in the study.

Results and Discussion

A total of 413 participants were recruited, with the mean age being 28.3 ± 7.9. Most of the participants were single (80.2%) and 50.9% of them were between the ages of 17-25 years. The majority of the participants were recruited from St George (39.8%) and the least (5.0%) from Carriacou (Table 1). Forty-four percent (44%) of the participants completed secondary school; 65% were employed; 14.4% were students; 10.3% stayed at home or were unemployed; and, 10.8% were self-employed. Of those persons who stayed at home or were unemployed, 51.2% were between the ages of 17-25 years old. Most of the participants indicated that they were sexually active (82.2%), and 59.8% of them had unprotected sex in the previous 30 days (Table 1). More than half of the participants had been tested for HIV and STIs (Sexually Transmitted Infections) (74.6% and 65% respectively), the differences being statistically significant ($p < 0.001$).

Table 1. Distribution of the socio-demographic characteristics of participants

Variable	**Age Groups**				**P-value**
	17 - 25 (%)	26 - 35 (%)	36 - 45 (%)	Total (%)*	
Parish					
St. Andrew	26 (46.4)	21 (37.5)	9 (16.1)	56 (14)	0.439
St. David	17 (50.0)	10 (29.4)	7 (20.6)	34 (8.5)	
St. George	63 (39.6)	57 (35.8)	39 (24.5)	159 (39.8)	
St. John	22 (39.3)	19 (33.9)	15 (26.8)	56 (14.0)	
St. Mark	10 (33.3)	15 (50.0)	5 (16.7)	30 (7.5)	
St. Patrick	24 (53.3)	16 (35.6)	5 (11.1)	45 (11.3)	
Carriacou	11 (55.0)	4 (20.0)	5 (25.0)	20 (5.0)	
Marital Status					
Single	165 (50.9)	111 (34.3)	48 (14.8)	324 (80.2)	< 0.001
Married	6 (10.0)	24 (40.0)	30 (50.0)	60 (14.9)	
Divorced	1 (12.5)	0 (0.0)	7 (87.5)	8 (2.0)	
Separated	1 (25.0)	2 (50.0)	1 (25.)	4 (1.0)	
Other	1 (12.5)	7 (87.5)	0 (0.0)	8 (2.0)	
Education					
Primary	5 (14.3)	10 (28.6)	20 (57.1)	35 (8.7)	< 0.001
Secondary	88 (48.9)	55 (30.6)	37 (20.6)	180 (44.7)	
Some college	23 (44.2)	22 (42.3)	7 (13.5)	52 (12.9)	
College graduate	56 (46.7)	47 (39.2)	17 (14.2)	120 (29.8)	

Table 1. (Continued)

Variable	**Age Groups**				**P-value**
	17 - 25 (%)	26 - 35 (%)	36 - 45 (%)	Total (%)*	
Post graduate	1 (6.3)	11 (68.8)	4 (25.0)	16 (4.0)	< 0.001
Occupation					
Student	53 (93.0)	3 (5.3)	1 (1.8)	57 (14.4)	< 0.001
Employed by Others	84 (32.8)	111 (43.4)	61 (23.8)	256 (64.5)	
Stay at home	21 (51.2)	14 (34.1)	6 (14.6)	41 (10.3)	
Self employed	11 (25.6)	17 (39.5)	15 (34.9)	43 (10.8)	
Living Status					
Own a house	126 (46.8)	84 (31.2)	59 (21.9)	269 (65.3)	0.038
Rent a house	21 (29.6)	35 (49.3)	15 (21.1)	71 (17.2)	
Squat	1 (10.0)	5 (50.0)	4 (40.0)	10 (2.4)	
Living with parents	15 (51.7)	11 (37.9)	3 (10.3)	29 (7.0)	
Other	14 (42.4)	13 (39.4)	6 (18.2)	33 (8.0)	
Sexually Active (Yes)	132 (40.2)	128 (39.0)	68 (20.7)	382 (82.2)	0.001
Unprotected Sex (Yes)	79 (39.1)	84 (41.6)	39 (19.3)	202 (59.8)	0.332

*Percentages based on the total number of responses with in each variable choice. (n = 400 for Parish, n = 404 for Marital Status, n = 412 for Living Status, n = 403 for Education, n = 397 for Occupation, n = 399 for Sexually active, and n = 338 for Unprotected sex.

Table 2. Distribution of health risk behavior by demographics

Variable	Transactional Sex		Condom Use		Non-regular sex partner		Unprotected sex	
	N (%)	p-value	N (%)	p-value	N (%)	p-value	N (%)	p-value
Parish	15 (4.4)	0.017	70 (20.8)	0.061	87 (26.8)	0.028	199 (59.6)	0.008
St. Andrew	2 (13.3)		9 (12.9)		22 (25.3)		38 (19.1)	
St. David	1 (6.7)		6 (8.6)		5 (5.7)		22 (11.1)	
St. George	3 (20.0)		22 (31.4)		26 (29.9)		78 (39.2)	
St. John	7 (46.7)		11 (15.7)		15 (17.2)		25 (12.6)	
St. Mark	1 (6.7)		3 (4.3)		10 (11.5)		14 (7.0)	
St. Patrick	1 (6.7)		12 (17.1)		6 (6.9)		19 (9.5)	
Carriacou	0 (0.0)		7 (10.0)		3 (3.4)		3 (0.9)	
Total Responses	341 (100)		336 (100)		325 (100)		334 (100)	
Education	15 (4.4)	0.008	70 (20.6)	0.137	89 (27.1)	< 0.001	202 (59.9)	0.035
Primary	5 (33.3)		10 (14.3)		9 (10.1)		12 (5.9)	
Secondary	3 (20.0)		35 (50.0)		36 (40.4)		86 (42.6)	
Some college	3 (20.0)		6 (8.6)		14 (15.7)		30 (14.9)	
College graduate	3 (20.0)		18 (25.7)		26 (29.2)		59 (29.2)	

Table 2. (Continued)

Variable	Transactional Sex		Condom Use		Non-regular sex partner		Unprotected sex	
	N (%)	p-value	N (%)	p-value	N (%)	p-value	N (%)	p-value
Post graduate	1 (6.7)	0.008	1 (1.4)	0.137	4 (4.5)	< 0.001	15 (7.4)	0.035
Total Responses	344 (100)		339 (100)		328 (100)		337 (100)	
Occupation	15 (4.4)	0.001	67 (20.0)	0.108	87 (26.9)	0.091	202 (60.7)	0.616
Student	2 (13.3)		11 (16.4)		15 (17.2)		24 (11.9)	
Employed by Others	4 (26.7)		36 (53.7)		48 (55.2)		138 (68.3)	
Stay at home/Unemployed	6 (40.0)		8 (11.9)		13 (14.9)		19 (9.4)	
Self employed	3 (20.0)		12 (17.9)		11 (12.6)		21 (10.4)	
Total Responses	340 (100)		335 (100)		324 (100)		333 (100)	

Behavioral Risk Factors

Transactional Sex

Transactional sex was reported by 4.4% of the participants; and, it was higher in women who only completed primary school (33.3%; p = 0.008), and those who were unemployed (40%; p = 0.001). These results are consistent with other studies' findings where transactional sex was associated with lower educational and economic statuses [11, 15]. The parish St John had the highest reported transactional sex (46.7%; p = 0.017) (Table 2). This parish is considered one of the poorest parishes in Grenada. However, there is limited information on the possible triggers in the parish and further investigations are needed. In addition to economic reasons, forced sexual initiation and being in an abusive relationship are also commonly associated with transactional sex [18, 19]; although the confidence interval (CI: 2.06 – 19.33) in the current study is broad, the findings still indicate a significant association between forced sexual initiation and transactional sex.

Unprotected Sex and Condom Use

It is alarming that a significant number of respondents reported having unprotected sex in the previous month (58.9%), and 48.8% of them have never been tested for HIV. Additionally, consistent condom use was only reported by 20.8% of the participants. These findings were similar to what has been observed in Barbados where 45.8% of persons aged 15-29 years old indicated inconsistent condom use, and 21.9% indicated that they never used condoms [13]. This could be due a low understanding of the risk, being in a committed relationship, or due to low condom negotiation abilities. Most often, the use of condoms by women in long-term relationships depends on the nature of the relationship they are in and the type of person with whom they are involved [20].

Condom use was lowest in the parish St Mark (4.3%). This parish is the most impoverished in Grenada with a poverty severity index of 9.3, which may account for the low condom use [21]. Moreover, compared with women from St. Andrew, women from St. Patrick and Carriacou were more likely to use a condom every time (OR = 4.43, 95% CI: 1.30 – 15.1, p = 0.017 and OR = 5.44, 95% CI: 1.21 – 24.5, p = 0.028 respectively); however, women from St Mark were less likely to use a condom with other partners (OR = 0.094, 95% CI: 0.011 – 0.81, p = 0.031). No other significant differences were noted.

When educational status was considered, a college graduate was less likely to use a condom every time compared with a woman having a primary school education (OR = 0.295, 95% CI: 0.088 – 0.99, p = 0.049). A woman with secondary school education (OR = 0.281, 95% CI: 0.097 – 0.82, p = 0.02) and college graduates (OR = 0.263, 95% CI: 0.083 – 0.84, p = 0.024) were less likely to use a condom at the last sexual intercourse compared with a woman having a primary school education. These finding are consistent with the results of a study of US college students (years 1-4) where it was found that the odds of condom use decreased by 40% from first year of college to fourth year of college [22]. This could be due to the perception that their partners are safe, or the perception that STIs are problems associated with persons from lower social levels or low educational backgrounds [25].

Socio-Cultural Factors

Societal Ideologies

Social influences, beliefs and exposure to certain environments also have an effect on health risk behaviors. Age at first sex, participants' belief of the appropriate age for first sex, and whom they think young girls should have sex with during their sexual initiation, were assessed. In the current study, the average age at which girls started having sex was 17 years old (± 2.8). This finding is different from the results of the Barbadian study [23], where 50% of the participants indicated that they had their first intercourse before 13 years of age. The difference may be due to bias in the self-reporting of age. However, there was no difference in reporting of the age of first sex by participants who completed the paper vs.

electronic questionnaire, which indicates that the sensitive nature of the questions did not lead participants to give the perceived socially accepted age (higher number).

Participants indicated that 18 (± 2.7) is the appropriate age at which a young girl could start having intercourse. This finding is higher than the 15-16 year age range noted researchers in Barbados [13]. The difference in age may be due to the study population differences where the researchers included females aged 14-16 [13], which is younger than the current study sample (17 – 45). In this study, there was a weak positive significant correlation between the participant's age and age of first sex ($r = 0.201$, $p < 0.001$), as well as between age and appropriate age to have sex ($r = 0.148$, $p = 0.007$); and age of first sex and appropriate age to have sex ($r = 0.332$, $p < 0.001$).

Forced first sexual intercourse (childhood sexual trauma) was reported by 12.1% of the participants. Similar to other studies, childhood sexual trauma was associated with sexual risk behaviors, especially transactional sex [8, 10]. In this study, 43% of the women who have exchanged sex for money experienced childhood sexual trauma ($p = 0.003$), and had higher odds of engaging in transactional sex (OR = 6.3; 95% CI: 2.06 – 19.33).

Furthermore, participants were asked if they would tolerate their partner having another woman. Of those persons responding "Yes," significant associations were noted for persons who exchanged sex for money ($p = 0.012$); those who had unprotected sex ($p < 0.001$); and those who were ever diagnosed with an STI ($p < 0.001$). Women who tolerated their partner having another woman had higher olds of engaging in transactional sex (OR = 6.12; 95% CI: 1.76 – 21.24), and being diagnosed with STIs (OR = 4.7, 95%CI: 1.66 – 13.34). In addition, 48.9% of those who had sex with a non-regular partner indicated that they agreed with the statement "In Grenadian culture manhood is associated with a man's ability to support more than one woman (meaning of manhood)," the difference being statistically significant ($p = 0.01$). These women were also found to be at increased odds of having sex with a non-regular partner (OR = 2.36, 95% CI: 1.30 – 4.29). These responses are indicative of a larger societal and community level issues [8].

Such a societal norm would have to be addressed in order to reduce the health impact of the associated risk factors.

Participants were also asked their opinion on who they thought young girls have sex with during their first intercourse. The majority (41.5%) of the participants from St George indicated "friend" or "boyfriend," while 28% of those from St John indicated "family member." The later finding is alarming since incest is known to occur; however, it is a subject that is taboo in the Caribbean region. This finding may be due to the personal experience of those women who indicated a family member.

Influence of Regional Music Genres

Participants were asked if they listened to Dancehall, Reggie or Soca music genres, and if their sexual behavior was influenced by the genres. Music influenced those who have exchanged sex for money ($p = 0.033$) and those who have reported having a sexual relationship with a non-regular partner ($p = 0.003$). Of those who exchanged sex for money, 66.7% indicated that their sexual behaviors were influenced by the music ($p = 0.033$), 64.3% indicated that they acted on the lyrics ($p = 0.001$) and 53.3% indicated that the music increased their desire for sex ($p = 0.002$).

In addition, 30.7% of those who indicated that they had sex with a non-regular partner reported that the music types increased their desire for sex ($p = 0.004$). A statistically significant association ($p = 0.003$) was also found between having a non-regular partner and the perception that the music types indicated that the best way to keep a man is by satisfying his sexual needs.

Of those who reported having sexual relations with a non-regular partner, 67.4% reported hearing the lyrics of the songs in their head ($p < 0.001$); 42.7% reported feeling the lyrics in their emotions ($p < 0.001$); 68.5% sang the lyrics without hearing them ($p = 0.021$); 31.5% reported that the lyrics inspired their dreams ($p = 0.002$); and 13.5% indicated that they acted on the lyrical dreams ($p = 0.003$). Statistically significant associations were also found when considering transactional sex and reported feeling the lyrics in their emotions ($p = 0.004$), as well lyrics inspiring dreams (0.011).

The influence of these genres and their lyrics could be because females are more likely to be impacted "psycho-socially" and are more likely to act on the lyrical contents [24]. A study conducted in Jamaica noted that females were more "emotionally stimulated" by Dancehall musical content compared to males (82% females vs 62% males) [14], and in another study the majority of the women reported that dancehall music made them want to have sex [24]. The dancehall music culture defines masculinity by the sexual exploitation and conquering of women [24], the music promoting the concept that a woman can keep her man by satisfying his sexual needs. The results of this study support this statement, as 29.7% of the women believed in this concept.

Economic Factors

Economically motivated relationships and sexual risk behaviors were assessed. Women who had multiple sexual partners (three or more) reported engaging in economically motivated relationships. Of those women 35.9% stayed in a relationship longer than they wanted to, 38% started a new relationship, and 31% had sex with a non-regular partner to get money for things they could not afford, the differences being statistically significant ($p < 0.001$). Some women get involved in economically motivated relationships with multiple partners to receive material support from one partner while obtaining emotional support from the other partner(s) [15].

Women involved in transactional sex also reported engaging in economically motivated relationships. Of these women, 73.3% stayed in a relationship longer than they wanted to; 66.7% started a new relationship; and 86.7% had sex with a non-regular partner to secure funds for buying things they could not afford or to support their children and other dependents, the differences being statistically significant ($p < 0.001$). Engaging in transactional sex for these reasons is a "manifestation of coping with living in an area characterized by high unemployment and poverty" [8].

Economic factors also influenced condom use. Of the participants who reported no condom use (at all) 20.4% ($p = 0.004$) stayed in a relationship longer than they wanted to and 17.8% started a new relationship ($p < 0.001$) for economic reasons. Additionally, 11.5% had unprotected sex with a non-regular partner to be able to pay for things they could not afford ($p = 0.004$) or to support a child (dependent) (19.6%, $p < 0.024$). This is consistent with other published studies that indicate minimal condom use if women expect any kind of assistance from their partners [15, 25]; highlighting the power imbalance within these relationships [26]. Another study from Kenya showed that women's request for condoms lowered the amount of support they get from their partners [27]; while several studies from South Africa noted that women were unsuccessful in requesting condoms from transactional partners [15, 28, 29].

Economics also influenced women who were involved with partners who injected drugs. Of those participants who were involved with partners who injected drugs, 66.7% stayed in the relationship because they needed help to pay their bills ($p = 0.001$); and 77.8% needed support to take care of their children ($p < 0.001$).

Intimate Partner Violence

Intimate partner violence in the forms of physical, psychological, and sexual abuse, and controlling behaviors was assessed. Participants who had multiple concurrent partners were subjected to physical abuse (34.5%; $p < 0.001$); psychological abuse (45.1%; $p = 0.002$); controlling relationships (45.8%; $p < 0.001$); and 52.8% were in an accusatory relationship ($p < 0.001$). Researchers indicate that women in abusive relationships involve with multiple concurrent partners because they often seek additional partners for emotional support, solace or as an act of resistance [30].

Also, women engaged in transactional sex were subject to psychologically abuse (66.7%; $p < 0.001$), physical abuse (53.3%; $p < 0.001$); controlling relationships (73.3%; $p < 0.001$), and accusatory relationships (73.3% $p = 0.004$). These women lack the power to advocate for themselves and the resource to help them escape these livelihoods [8].

Social Constructionism Concept

The findings of this study can be explained using the Social Constructionism Concept, in which "knowledge and truth are created by interaction of individuals within society" [31]. Health behaviors are used in daily interactions in the society [32] and society gives meaning to behavior and it becomes norm. Risks are therefore experienced in the social and cultural settings; and culture influences individual thoughts and behavior [13]. Also, factors such as educational level, economic status, and social context influence the kind of behavior people construct and contribute to different health risks [32]. Participants in the current study live in a risk-charged environment (low income, low employment opportunities, low education level). The theoretical perspective of social constructionism can therefore allow for a better interpretation of health risk behaviors in Grenadian women, and how they experience and negotiate their realities and risks.

Limitations

The most significant limitation of this study was the sensitive topic of study, which could have introduced response bias to our findings. Participants were given the option of completing the survey themselves, and to answer only what they feel comfortable answering. However, those with lower educational levels needed interviewer assistance in completing the questionnaires. This could have caused participants to give responses based on what they perceived the interviewer would expect to hear due to socially accepted norms. Also, the length of the survey (65 questions) restricted recruitment, as only those who had the time and energy to go through all the questions were able to participate. The sampling strategy could have also introduced selection bias. In addition, females younger than 17 years old were not included; therefore, these findings cannot be generalizable to these age groups.

Conclusion

Assessing and understanding the underlying risk factors of HIV is the first step in preventing and controlling the spread of the virus. This study is the first to investigate the socio-behavioral risk factors of HIV for Grenadian women. The major behavioral risk factors identified in this study were inconsistent condom use/unprotected sex, having multiple partners and transactional sex. These behaviors were influenced by societal ideologies and norms; and in some cases were exacerbated by exposure to intimate partner violence and childhood sexual trauma. Based on the social constructionism theory, addressing the issues at the societal and institutional levels would be more effective. However, immediate actions should be taken by educating and providing resources to the at risk population at the high-risk places.

Competing Interests

The authors declare that they have no competing interests.

Authors' Contributions

WG played key roles of formulating the project, designing the study, collecting data in the field, analyzing and interpreting data and writing the manuscript. SR provided advice and corrections, participated in the design of the study, data analysis, and general supervision and provision of advice to ensure proper development of the work. All authors read and approved the final manuscript.

Acknowledgments

We would like to thank the St. George's University, School of Medicine and the Department of Public Health and Preventive Medicine, St George Grenada. Ms. Shantel Peters-St. John for her support in reviewing the project study design. Dr. Christine Richards for her support in assistance during data collection and advice in identifying a theory to describe results. Ms. Tashi Creft, Mr. Etieron Edwards, Dr. Birtukan Ejigu, Ms. Dianne Roberts, Ms. Ozioma Onyegbule for their help with data collection.

References

[1] WHO | *HIV/AIDS. Media centre Fact sheets*. 2014.
[2] *Global Statistics*. AIDSgov. 2014.
[3] Padilla MB, Guilamo-Ramos V, Bouris A, Reyes AM. HIV/AIDS and tourism in the Caribbean: an ecological systems perspective. *American journal of public health.* 2010;100:70-7.
[4] *Caribbean HIV & AIDS Statistics*. AVERT. 2014.
[5] (MOH) GMoH, editor Grenada *HIV/AIDS Surveillance Data 2012*; St George.
[6] Lemme F, Doyle AM, Changalucha J, Andreasen A, Baisley K, Maganja K, et al. HIV Infection among Young People in Northwest Tanzania: The Role of Biological, Behavioural and Socio-Demographic Risk Factors. *PLoS One*. 2013;8(6):e66287.
[7] *AIDS in the Caribbean - Statistics, History, and Policies* | Medwiser. 2014.
[8] Frew PM, Parker K, Vo L, Haley D, O'Leary A, Diallo DD, et al. Socioecological factors influencing women's HIV risk in the United

States: qualitative findings from the women's HIV SeroIncidence study (HPTN 064). *BMC Public Health.* 2016;16(1):803.

[9] Quinn TC, Overbaugh J. HIV/AIDS in women: an expanding epidemic. *Science* (New York, NY). 2005;308:1582-3.

[10] Hahm HC, Kolaczyk E, Lee Y, Jang J, Ng L. Do Asian-American women who were maltreated as children have a higher likelihood for HIV risk behaviors and adverse mental health outcomes? *Women's health issues: official publication of the Jacobs Institute of Women's Health.* 2012;22(1):e35-43.

[11] Bombereau G, Allen C. Social and cultural factors driving the HIV epidemic in the Caribbean: a literature review. *Caribbean Health Research Council.* 2008.

[12] Versteeg M, Murray M. Condom use as part of the wider HIV prevention strategy: Experiences from communities in the North West Province, South Africa. *SAHARA-J: Journal of Social Aspects of HIV/AIDS.* 2008;5:83-93.

[13] Barrow C. *Adolescent Girls, Sexual Culture, Risk and HIV in Barbados*. 2007. p. 1-15.

[14] Crawford AD. The effects of dancehall genre on adolescent sexual and violent behavior in Jamaica: A public health concern. *North American journal of medical sciences.* 2010;2:143-5.

[15] Dunkle KL, Wingood GM, Camp CM, DiClemente RJ. Economically motivated relationships and transactional sex among unmarried African American and white women: results from a U.S. national telephone survey. *Public health reports* (Washington, DC: 1974). 2010;125 Suppl 90-100.

[16] *Evaluation of a theatre performance for adolescents promoting safer sex behaviour using AIDS risk reduction theories*. (2009).

[17] Shepard MF, Campbell JA. The Abusive Behavior Inventory: A Measure of Psychological and Physical Abuse. *Journal of Interpersonal Violence.* 1992;7:291-305.

[18] Dunkle KL, Jewkes RK, Brown HC, Gray GE, McIntryre JA, Harlow SD. Gender-based violence, relationship power, and risk of HIV

infection in women attending antenatal clinics in South Africa. *Lancet* (London, England). 2004;363:1415-21.

[19] Stuart S. *The Reproductive Health Challenge: Women and AIDS in the Caribbean*. The Caribbean AIDS Epidemic: University of the West Indies Press; 2000. p. 122-38.

[20] Exavery A, Kanté AM, Jackson E, Noronha J, Sikustahili G, Tani K, et al. Role of condom negotiation on condom use among women of reproductive age in three districts in Tanzania. *BMC Public Health.* 2012;12:1097.

[21] *Final Report Country Poverty Assessment*: Grenada, Carriacou and Petit Martinique 2008.

[22] Bearak JM. Casual Contraception in Casual Sex: Life-Cycle Change in Undergraduates' Sexual Behavior in Hookups. *Social Forces.* 2014;93:483-513.

[23] Ellis H, Hoyos MD, Jones F, Roach TC, Souder M, Walrond ER. Knowledge, attitude, belief and practices survey in relation to AIDS amongst children, ages 11-16 years old in Barbados. *Bull E Carib Affairs.* 16:1-12.

[24] Hill S-m, editor *The Influence of Popular Culture on Male Sexual and Reproductive Health Behaviour*. 2010.

[25] *Sexual behavior and condom use in female students in Dar-es-salaam, Tanzania: Differences by steady and casual partners*. 2014.

[26] Hunter TS, Tilley DS. A Grounded Theory Study of the Process Used to Negotiate Condom Use among African-American Women: Review of the Literature. *Journal of cultural diversity.* 2015;22(1):23-9.

[27] Luke N. Exchange and Condom Use in Informal Sexual Relationships in Urban Kenya. *Economic Development and Cultural Change.* 2006;54:319-48.

[28] Hunter M. The Materiality of Everyday Sex: Thinking beyond 'prostitution'. *African Studies.* 2010;61:99-120.

[29] Silberschmidt M, Rasch V. Adolescent girls, illegal abortions and “sugar-daddies” in Dar es Salaam: vulnerable victims and active social agents. *Social Science & Medicine.* 2001;52:1815-26.

[30] Teitelman AM, Tennille J, Bohinski J, Jemmott LS, Jemmott JB. Urban adolescent girls' perspectives on multiple partners in the context of the sexual double standard and intimate partner violence. *The Journal of the Association of Nurses in AIDS Care: JANAC.* 2013;24:308-21.

[31] Andrews T. What is Social Constructionism? | Grounded Theory Review. *Grounded Theory Review.* 2012;11.

[32] Courtenay WH. Social Constructionism has been instrumental in remodeling grounded theory. In attempting to make sense of the social world, social constructionists view knowledge as constructed as opposed to created. This paper discusses how social constructionists cons. *Social Science and Medicine.* 2000;50:1385 - 401.

Date Submitted: June 7, 2017. *Date Accepted:* September 4, 2017.

In: Current STEM. Volume 1
Editor: Maurice HT Ling
ISBN: 978-1-53613-416-2

Chapter 3

A CRYPTOGRAPHY METHOD INSPIRED BY JIGSAW PUZZLES

Maurice HT Ling*
Colossus Technologies LLP, Singapore
School of BioSciences, The University of Melbourne, Australia

ABSTRACT

Cryptography is a critical tool in safeguarding information from "unauthorized" view during the storage and transportation of data. Due to the one-to-one correspondence between plain text and cipher text, encryption algorithms can be seen as a transformation process. This is a deficiency as all information is present, though encrypted, in the cipher text. Inspired by Jigsaw puzzles, a new cryptography system, Jigsaw Encryption System (JES) is proposed where a single plain text file results in many cipher text files, resembling jigsaw pieces from a single image; thus, the loss of a small number of cipher text files may not compromise the entire contents in plain text. Each cipher text can be further processed for added security. This can result in larger permutations needed to decipher by brute force. Reference implementations of preliminary JES versions had been deposited into COPADS (Collection of Python

* Corresponding Author Emails: mauriceling@acm.org, mauriceling@colossus-tech.com.

Algorithms and Data Structures) repository (https://github.com/mauriceling/copads; File name: copads/jigsaw.py).

Keywords: encryption for transportation, file splitting, Jigsaw, Python

INTRODUCTION

Cryptography is synonymous with code writing and has its roots dated to at least 1900 BC when Egyptians used non-standard hieroglyphs to conceal the meaning of their writings (Whitman and Mattord, 2005). Salomaa cleverly explained the two-world concept of cryptography – the sunlit world of secured information exchange, and the dark world of intercepting and decoding this information (Salomaa, 2013). The loss of secured information may have dire consequences, as exemplified in Babington Plot leading to the trial and public beheading of Mary Queen of Scots by her cousin, Queen Elizabeth of England, in 1587 (Lewis, 1999).

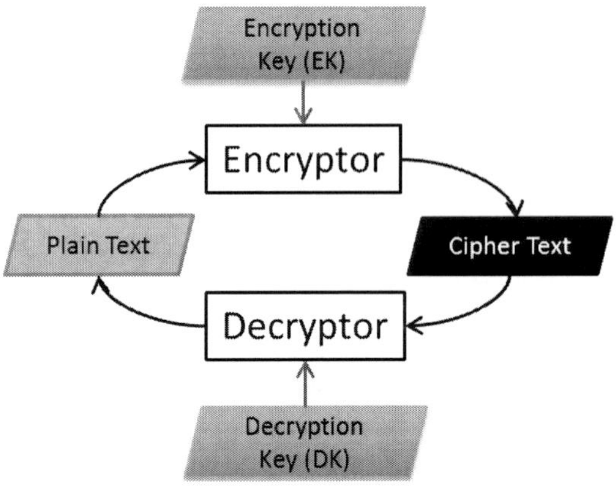

Figure 1. Encryption-Decryption Cycle. Unencrypted data is denoted as plain text, which is converted to encrypted data denoted as cipher text, using an encryptor. The encryptor requires an encryption key. The cipher text can then be decrypted by a decryptor, requiring a decryption key, and produces the plain text. For the cipher text to be correctly decrypted, the correct decryption key to the encryption key must be used.

There are four components to any cryptography system (Figure 1). To convert an unsecured and human-readable text (usually known as plain text) into a secured and human-unreadable text (usually known as cipher text), an encryption mechanism (known as encryptor) is needed. The encryptor takes the plain text and an encryption key as input to produce the cipher text. Conversion from cipher text back to plain text will require the cipher text and decryption key as input to the decryptor. Hence, the encryptor and decryptor for a cryptographic system exist as a pair, and the encryption key and decryption key exist as a pair. When the encryption key is the same as the decryption key, the cryptographic method is known as symmetric-key cryptography, such as AES (Daemen and Rijmen, 2013) and BlowFish (Schneier, 1994). However, the encryption key and decryption key are different in some cryptographic method. This is known as asymmetric-key cryptography or public-key cryptography, such as RSA (Rivest et al., 1978) and ElGamer (ElGamal, 1985). The strength of the system lies in the difficulty in obtaining the encryption-decryption key-pair (Schneier, 1994), which can be translated to the number of possible encryption-decryption key-pairs. Hence, the larger the possible number of key-pairs, the stronger the encryption (Schneier, 1994). As the number of possible key-pairs is proportional to the computing time required obtaining the correct key-pairs using brute force methods (Bhanot and Hans, 2015) and the number of possible key-pairs is largely correlated to the length of the key, the length of the key is commonly used to compare the strength of the cryptographic method (Bhanot and Hans, 2015; Mane, 2015). In addition, the length of keys is largely dependent on the nature of the cryptographic algorithm (Bhanot and Hans, 2015; Mane, 2015); hence, difficult to expand without crucial changes to the algorithm.

Another way to view the encryptor and decryptor is that they are data transposition algorithms between plain text and cipher text; where plain text can be any data stream (Jothi, 2013), including images (Bai and Wu, 2016). In this point of view, cipher text is essentially a transposed text containing all the information and data in the plain text. In addition, the number of possible key-pairs can represent the total number of ways in which a cipher text can be decrypted. Thus, the entire weight of the cryptographic method rests on the strength of the encryption-decryption key-pairs. In Babington Plot, Sir Francis Walsingham, Queen Elizabeth's secretary, knew of the plot to assassinate Queen Elizabeth for some time but lacked the crucial evidence to incriminate Mary Queen of Scots to the assassination plot; hence, unable to convince Queen Elizabeth of the plot; until a single letter from Mary Queen of Scots approving the plot that finally convinced Queen Elizabeth and sealed the fate for Mary Queen of Scots. Therefore, the possible weakest link in any cryptographic system may be the fact that all the information in the plain text is found within a single cipher text, inviting and awaiting deciphering. The same can be said for each intercepted message sent to Bletchley Park for decryption during World War II.

In this study, a new cryptographic method based on Jigsaw puzzle is proposed; hence, named as Jigsaw Encryption System (JES); to address the weakness of full information correspondence between a single plain text and a single cipher text. In addition, Jigsaw encryption has no inherent key length requirement, which allows for key length expansion with relative ease. Calculation of the number of possible means to revert the cipher text back to plain text, which corresponds to the possible number of key-pairs, demonstrates that Jigsaw cryptography outperforms current cryptographic systems.

Jigsaw Cryptography

Inspired by jigsaw puzzles to address the weakness of full information correspondence between a single plain text and a single cipher text, Jigsaw cryptography proposes to slice one plain text file into multiple cipher text files during encryption process. This is likened to taking a large picture image and chopping it up in many jigsaw pieces. Using the analogy of Babington plot, Jigsaw Encryption System (JES) will be similar to shredding Mary Queen of Scots' approval letter and send each labeled shred separately to Anthony Babington, the chosen assassin. In the case of Enigma encrypted message sent by Germany during World War II, multiple instructions for various combat units can be concatenated together and shredded by JES before transmission, which adds another layer of complexity as each intercepted shred will not contain the entire combat order.

In this study, three versions of JES with increasing complexity have been implemented. As the possible number of key-pairs (also known as permutations) is proportional to the strength of the encryption [6], the number of permutations needed decrypt a 1 megabyte file by brute force method will be emphasis, on basis that each Jigsaw-ed cipher file (hereinafter known as Jigsaw file) is set to 64 kilobytes.

JES version 1 (JES1) encryptor takes a plain text file and split the file into a set of Jigsaw files based on a specified size for each Jigsaw file (Figure 2), and generates a log file. There are 2 modes for each JES version, even slicing and uneven slicing. In even slicing, each Jigsaw file will be of the same length as specified except the last Jigsaw file. In uneven slicing, the size of each Jigsaw file will be 1 to 2 times of the specified length. The generated log file will be used as input, together with the set of Jigsaw files, to the decryptor for assembly into the original plain text file. To ensure fidelity, file hashes for the original and assembled plain text file, as well as for each Jigsaw file, are generated and compared. Assuming that JES1 is used to encrypt a 1 MB (1,048,576 bytes) file and split into 16 64-KB Jigsaw files, there are 2 x 10^{13} permutations to assemble 16 64-KB files into the original 1 MB file.

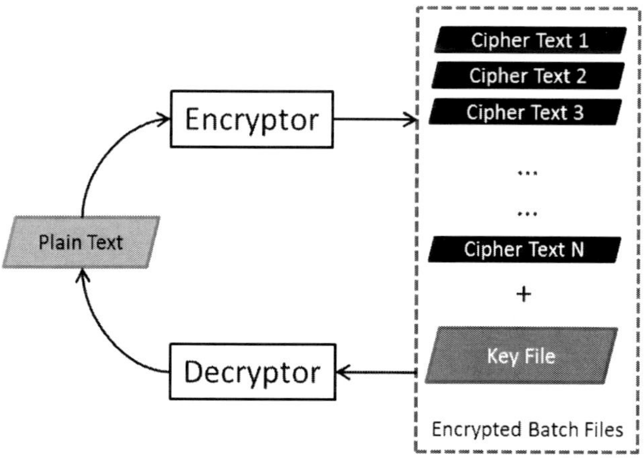

Figure 2. Schema of Jigsaw Cryptography.

However, JES1 cannot be considered to be true cryptographic method as each Jigsaw file are essentially unencrypted sections of the original plain text. This is likened to cutting a book into pages, remove all page numbers, and jumbling the pages up. The cipher text files are really sectioned plain text files. Despite so, the complexity in reassembly can be exponentially complex in event whereby the number of plain text files increases – instead of splitting a book into pages, the entire library of books is now split into pages.

JES version 2 (JES2) builds on JES1 by provides 63 randomly assigned variations for each Jigsaw file. Each variation is the reversion of a portion of the Jigsaw file (Figure 3). The difference between each reversion scheme is the start and end byte positions of the inversion. Bytes in between the start and stop positions are reversed. As a result, each Jigsaw file is no longer sectioned plain text as in JES1. Including the permutations on each JES1 Jigsaw file, the full permutations possible for 16 JES2 64-KB Jigsaw files is $(2 \times 10^{13})^{63}$ or 9×10^{837}.

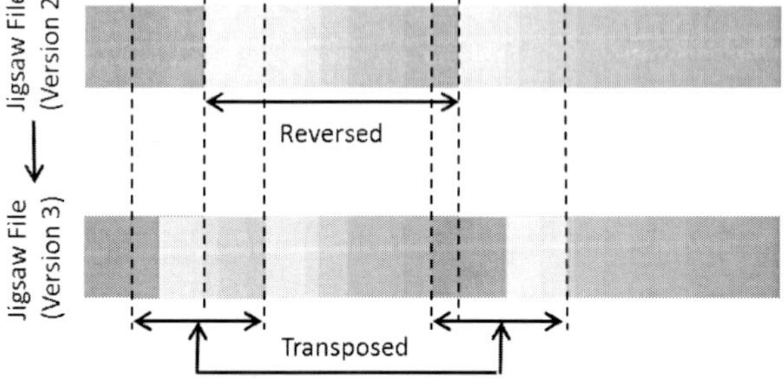

Figure 3. Relationship between Jigsaw Files of 3 Different Versions of Encryption.

JES version 3 (JES3) adds an additional step of transposition from the Jigsaw file provided by JES2 by applying one of the 1,350 randomly assigned transpositions to each Jigsaw file (Figure 3). The transposition function considers a Jigsaw file into two equal halves before randomly assigning one of the 1,350 start position and length in the first half of the Jigsaw file. It then takes the bytes as identified in the first half of the Jigsaw file and transpose it with the equal number of bytes extracted from the second half of the Jigsaw file starting from the beginning of the second half (the midpoint of the Jigsaw file). As a result, there can be a section of the Jigsaw file that is both reversed and transposed, adding to the complexity for deciphering. This means that there are possibly $63^{1,350}$ or $5 \times 10^{2,428}$ permutations for each of the 16 JES3 64-KB Jigsaw files generated for a 1 MB file, Hence, the full permutations possible for 16 JES3 64-KB Jigsaw files is $(9 \times 10^{837})^{1,350}$ or $6 \times 10^{1,131,240}$.

IMPLEMENTATION AND TESTING

JES versions 1 to 3 have been implemented in Python programming language and deposited in COPADS (Collection of Python Algorithms and Data Structures) repository (https://github.com/mauriceling/copads; File name: copads/jigsaw.py) under Python Software Foundation License version 2. Besides setting the version of JES to use and the size of Jigsaw file (as blocksize), the following operational parameters are available:

- Type of slicer (`setting = 'slicer'`), where "even" means that each Jigsaw file will be of the same size except the final Jigsaw file, and "uneven" means that each Jigsaw file size will be between block size and twice of block size
- Length of Jigsaw file name (`setting = 'filename length'`), and it is usually better to set a longer file name to prevent name collision
- Hash length of each Jigsaw file (`setting = 'hashlength'`), which will be generated for each Jigsaw file for debugging purposes

The following code exemplifies using JES to encrypt and decrypt a MP4 file:

```
import jigsaw
j = jigsaw.JigsawFile()
j.setting('slicer', 'even')
j.setting('blocksize', 32768) # 32KB Jigsaw file
j.setting('filenamelength', 30)
j.setting('hashlength', 16)
j.setting('version', 3)
j.setting('verbose', 2)

# Encrypt file: A.mp4
keyFileName = j.encrypt('A.mp4')

# Decrypt file to A_1.mp4
j.decrypt(keyFileName, 'A_1.mp4')
```

During encryption, the following summary will be generated.

```
Encrypting file: A.mp4
... Sorting out input (source file) directory
...... Attempt to get input directory from input
file path
...... Failed: Input file is relative file name
...... Set input directory to currect working
directory
... Sorting out output (encrypted files output)
directory
...... Output directory not given
...... Attempt to find output directory from input
file
...... Failed: Input file is relative file name
...... Set output directory to currect working
directory
... in input directory: C:\Users\Maurice
Ling\Desktop\jigsaw
... onto output directory: C:\Users\Maurice
Ling\Desktop\jigsaw
```

```
... using Jigsaw version: JigsawFileTHREE
... using slicer: even
... using block size: 32768
Writing key file: C:\Users\Maurice
Ling\Desktop\jigsaw\A.mp4.jgk
Processing using even slicer
0 blocks processed
1000 blocks processed
2000 blocks processed
3000 blocks processed
3920 blocks processed
Encrypting file, A.mp4, completed
```

The key file, which represents the decryption key, is generated with the file name format as `<plain text file name>.jgk`. In this case, the key file is `A.mp4.jgk` as the plain text file is `A.mp4`. This key file is required for decryption as it contains the assembly scheme. During decryption, the following summary will be generated. A series of file hashes will be generated from the decrypted file and compared with the corresponding file hashes generated from the original plain text file during encryption process and stored in the key file.

```
Decrypting file using keyfile: C:\Users\Maurice
Ling\Desktop\jigsaw\A.mp4.jgk
... Processing key file
... Sorting out input (encrypted files) directory
...... Input directory is not given
...... Attempt to get input directory from keyfile name
...... Keyfile name is absolute path: C:\Users\Maurice
Ling\Desktop\jigsaw\A.mp4.jgk
...... Set input directory to directory of keyfile
... Sorting out output file (unencrypted file) name
...... Given output file name is relative path: A_1.mp4
...... Attempt to generate absolute output file path
...... Set folder to write output file as input directory
```

```
... Directory of encrypted files (input): C:\Users\Maurice
Ling\Desktop\jigsaw
... Uncrypted file name (output): C:\Users\Maurice
Ling\Desktop\jigsaw\A_1.mp4
Decrypting file ......
0 blocks processed
1000 blocks processed
2000 blocks processed
3000 blocks processed
3920 Jigsaw files processed
Expected number of bytes: 128388534
Actual number of bytes : 128388534
File Hashs (Decrypted File vs Original Unencrypted File)
md5: 49c978e1e1124a3eb245eb7430843152
 vs 49c978e1e1124a3eb245eb7430843152
sha1: 43355d06548dee9027afaf4825b1f82226eb298e
 vs 43355d06548dee9027afaf4825b1f82226eb298e
sha224:
      54ea7a0e419ce7d28a68514815ceab811131b4923f671973dbaa
      119c
 vs
      54ea7a0e419ce7d28a68514815ceab811131b4923f671973dbaa
      119c
sha256:
      d2dc6e12f4f9f7d7ccf4a8e1ea41566a1a8041c3ad0863194daa
      081d2507d068
 vs
      d2dc6e12f4f9f7d7ccf4a8e1ea41566a1a8041c3ad0863194daa
      081d2507d068
sha384:
      17f9e4fbddb316ed3c87ed83afd1db31f73bb351d4d48e3af17e
      21b2f13fe9aa6de3bbe2d4c36ef47dad92058fde2926
 vs
      17f9e4fbddb316ed3c87ed83afd1db31f73bb351d4d48e3af17e
      21b2f13fe9aa6de3bbe2d4c36ef47dad92058fde2926
sha512:
      b0d64e8a12c25103de65562d22e38a8a89a6b95be38471bcf376
      a613e9e158abb97ee3f87848192b35ee005434713a4e712db235
      3912f091b4835478b80d095b
```

```
vs
      b0d64e8a12c25103de65562d22e38a8a89a6b95be38471bcf376
      a613e9e158abb97ee3f87848192b35ee005434713a4e712db235
      3912f091b4835478b80d095b
Decryption completed
```

In the example above, all 6 file hashes (md5, sha1, sha224, sha 256, sha384, and sha512) are identical for both before (original plain text) and after (decrypted cipher text) an encryption-decryption cycle, indicating that the JES encryptor and decryptor are functioning as expected.

DISCUSSION

Cryptographic methods are integral to information security during storage and transportation. However, one-plain text/one-cipher text relationship may represent a weak link in most cryptographic systems as all the information in the plain text exist in the cipher text, though encrypted. This study proposes to address this weakness by a 1 to N relationship between plain text and cipher text. As a result, the interception of a small number of cipher texts is unlikely to compromise the security of the plain text.

256-bit Advanced Encryption Standard (AES-256) is currently considered resistant to attacks [13] and suitable for use in many encryption applications [14]. Given that the key size of AES-256 is 32 characters long (32 characters of 8 bytes each = 256 bytes) and given that there are 94 characters in the set of mixed alphanumeric with symbols (that is, abcdefghijklmnopqrstuvwxyzABCDEFGHIJKLMNOPQRSTUVWX YZ0123456789!@#$%^&*()-_+=~`[]{}| \:;"'<>,.?/), the total number of possible keys is 94^{32} or 1.38×10^{63}. In comparison, the number of permutations for 16 JES1 Jigsaw files is significantly lesser than the total number of possible keys in AES-256 but the number of permutations for 16 JES2 or JES3 Jigsaw files are many magnitudes larger than that of AES-256. It has to be noted that the calculations are based on a

1 megabyte plain text file. The number of 64-KB Jigsaw files will increase linearly with large plain text files, leading to exponential increase in assembly complexity in terms of permutations. Moreover, the number of permutations for JES1 can be easily increased by reducing the size of each Jigsaw file, such as from 64-KB to 8-KB will increase the number of Jigsaw files by 8 times.

Three versions of JES, with increasing complexity, are presented in this study as example implementations; demonstrating the relative ease in extending this concept into increasingly complex cryptographic systems in the future. For example, it may be possible to encrypt sections of the plain text files with existing cryptographic systems, such as randomly choosing AES [5] or BlowFish [6] for each section, before passing through JES. In JES2, only one reversal is carried out; and JES3, only one reversal and one transposition are carried out. Future versions may include multiple rounds of reversals or/and transpositions. Other possible enhancements may include cross-encryption where 2 or more plain text files are encrypted together, or the inclusion of decoys where short sections of novels may be interspaced at random into a business document before encryption.

The main feature of JES is the 1-to-M plain text/cipher text relationship. Although this will allow secured transportation of data by separating the set of Jigsaw files into separate storage devices for transporting, this also meant that JES requires a large volume of disk I/O due to file writing and reading, resulting in significant processing time. Hence, JES may only be suitable for securing data archives and transportation rather than real-time use (Xie et al., 2016), such as in filesystem-level encryption [16], where encryption/decryption efficiency is critical.

References

Bai, K, and Wu, C. (2016). An AES-Like Cipher and Its White-Box Implementation. *The Computer Journal 59*, 1054–1065.

Bhanot, R, and Hans, R. (2015). A review and comparative analysis of various encryption algorithms. *International Journal of Security and Its Applications 9*, 289–306.

Daemen, J, and Rijmen, V. (2013). *The design of Rijndael: AES-the advanced encryption standard* Springer Science & Business Media.

ElGamal, T. (1985). A public key cryptosystem and a signature scheme based on discrete logarithms. *IEEE Transactions on Information Theory 31*, 469–472.

Jothi, L. (2013). A Literature Review: Cryptography Algorithms for Wireless sensor networks. In *International Journal of Computer Science and Information Technologies*, p.

Lewis, JE. (1999). *The trial of Mary Queen of Scots: A Brief history with documents* Macmillan.

Mane, R. (2015). A Review on Cryptography Algorithms, Attacks and Encryption Tools. *International Journal of Innovative Research in Computer and Communication Engineering 3*, 8509–8514.

Rivest, RL, Shamir, A, and Adleman, L. (1978). A method for obtaining digital signatures and public-key cryptosystems. *Communications of the ACM 21*, 120–126.

Salomaa, A. (2013). *Public-key cryptography* Springer Science & Business Media.

Schneier, B. (1994). *Description of a new variable-length key, 64-bit block cipher (Blowfish)*. Springer, pp. 191–204.

Whitman, M, and Mattord, H. (2005). *Principles of information security.* Canada, Thomson Learning, Inc. *4*, 2009.

Xie, Y, Ding, W, and Wang, Y. (2016). A More Extensible Transparent Encrypted File System Based on Filter Driver. *Journal of Communications 11*, 383–387.

Date Submitted: August 4, 2017. *Date Accepted*: September 7, 2017.

In: Current STEM. Volume 1
Editor: Maurice HT Ling

ISBN: 978-1-53613-416-2

Chapter 4

COPADS V: LINDENMAYER SYSTEM WITH STOCHASTIC AND FUNCTION-BASED RULES

*Maurice HT Ling**
Colossus Technologies LLP, Singapore
School of BioSciences, The University of Melbourne, Australia

ABSTRACT

Lindenmayer system, commonly known as L-system, is a string rewriting system based on a set of rules. In each iteration, the string is repeatedly rewritten based on the rules given. This has been used to model branching processes; such as, plant and animal body patterning, and sedimentation processes. In addition to deterministic rewriting rules, stochastic rules have been used, leading to the development of stochastic L-system (S-L-system). For more complex modeling, parametric rules have been used, leading to the development of parametric L-system (P-L-system). Combining S-L-system and P-L-system leads to the development of L-system capable of handling both stochastic and parametric rules or parametric-stochastic-L-system (PS-L-system). Currently, there is no pure Python PS-L-system library. In this study, a light-weight, pure Python PS-L-system has been implemented. This has been incorporated into COPADS

* Corresponding Author Emails: mauriceling@acm.org, mauriceling@colossus-tech.com.

repository (https://github.com/copads/copads) and and licensed under GNU General Public License 3.

Keywords: rewriting rules, lindenmayer system, L-system, parametric-stochastic system, algorithmic beauty, python module

Introduction

Lindenmayer system, or commonly known as L-system, is invented by Aristid Lindenmayer (Lindenmayer, 1968a; Lindenmayer, 1968b) as a string rewriting system, originally meant to describe the development of multicellular organisms (Lindenmayer, 1971; Lindenmayer, 1975). The main advantage of L-system lies in its simplicity. For example, if every occurrence of "F" is to be rewritten into "FFRR" and given "F" as a starting string, the next 3 rewriting iterations will generate "FFRR", "FFRRFFRRRR," and "FFRRFFRRRRFFRRFF RRRRRR" respectively. Hence, complex strings can be generated from simple rewriting rules over a number of iteration, commonly known as generations. By coding "F" and "R" as "forward" and "turn 60° angle right" respectively using a LOGO-like turtle, complex structures may be formed. However, it has also been shown that L-systems are Turing complete (van Dalen, 1971); thus, all algorithmic structures are possible.

Shortly after the initial description of L-system (Lindenmayer, 1968a), Lindenmayer (1968b) explains how such a rewriting system can be model branching processes. Using this approach, L-systems have been used for the modeling of various plant structures (Cici et al., 2009; Room et al., 1996; Perttunen and Sievänen, 2005; Prince et al., 2014; Prusinkiewicz and Lindenmayer, 1990; Taralova et al., 2011), body plans of animals (Hoyal Cuthill and Conway Morris, 2014; Davoodi and Boozarjomehry, 2016), generation of 3-D computer graphics world (Fridenfalk, 2016) and soil sedimentation (Rongier et al., 2017a; Rongier et al., 2017b), in recent years.

The most basic form of L-system consists of only replacement rules where one instruction is replaced with one or more instructions. This can be known as replacement L-system (R-L-system). Given the rules and starting string (also known as axiom), the outcomes of an rL-system is deterministic. Non-determinism can be achieved by adding a stochastic parameter to each rule (Samal et al., 1994; Taralova et al., 2011), which gives a probability of executing a rewriting rulem, resulting in a stochastic L-system (S-L-system). S-L-systems allow for multiple rules identifying the same symbol to be executed at different probabilities. R-L-system is also a context-free system, as the execution of replacement rules is solely determined by the current instruction and not the context (what is before and after) of the instruction. Context sensitivity can be achieved by parameterizing the instruction to take account of the contextual landscape of the current instruction (Shi et al., 2011; Xin et al., 2014), resulting in a parametric L-system (P-L-system). Combining S-L-system with P-L-system results in a L-system, which can be both non-deterministic and context sensitive (Parish and Müller, 2011), that can be known as parametric-stochastic L-system (PS-L-system). Recently, a PS-L-system has been used to model blood vessels in organs (Galarreta-Valverde et al., 2013).

In terms of Python-related implementations, there are L-Py (Boudon et al., 2012) and lsystem by Erik Max Francis (http://www.alcyone.com/pyos/lsystem/). However, lsystem is a R-L-system, which does not allow for probabilistic or parametric rules. L-Py (Boudon et al., 2012) is implemented as a C++ library embedded in Python and focused on the flexibility to use any Python objects as instructions and parameters.

In this code manuscript, a pure Python, single file implementation of a PS-L-system, which focus of string rewriting in contrast to L-Py (Boudon et al., 2012), is presented. This implementation allows rules to be executed based on priorities, a feature not found in most (if not all) L-systems. This implementation has been incorporated into COPADS codebase (https://github.com/copads/copads) and licensed under GNU General Public License 3.

Code Description

The entire Lindenmayer system is implemented the system as a class (lindenmayer class in lindenmayer.py), which comprises of a constructor, three public methods (*add_rules*, *generate*, and *turtle_generate*), and three private methods. The constructor takes a parameter to indicate the length of each instruction or command to process, which is default at one. For example, given a string of AABBACCC and the instruction length is one, the re-writing rules will be executed using parameters in the following sequence – 'A', 'A', 'B', 'B', 'A', 'C', 'C', 'C'. However, if the instruction length is two, the re-writing rules will be executed using parameters in the following sequence – 'AA', 'BB', 'AC', 'CC'. This means that varying instruction lengths is not allowed.

Once the system is instantiated, the second step is to add rules, which are implemented as Python list of lists, into the PS-L-system using *add_rules* method. Rewriting rules are the crux of a L-system. There are three types of rules: (1) replacement rule, which replaces the current instruction for one or more instructions; (2) probability rule, which is essentially a replacement rule with less than or equal to 100% chance of execution; and (3) function rule, which calls a function and takes in the entire instruction string and the position of the current instruction. Function rules are used to implement parametric rules; and parametric, stochastic rules.

The simplest rule is a replacement rule without priority. The rule "substitute or rewrite X into XYY" can be written as a 2-element list of `['X', 'XYY']`. If this is the only rule, it has to be written as a list of lists, `[['X', 'XYY']]`, in order to be added into the system. If there are more than one rule in a set; each rule can have a different priority, which is added as the 3rd element in the rule list. For example, the rule set `[['X', 'XYY', 1], ['Y', 'XXY', 2]]`, will result all `'X'`s be rewritten as `'XYY'` before all `'Y'`s be rewritten into `'XXY'` as the execution order of the rules will be in ascending order of priorities. Hence, rule set `[['X', 'XYY']]` is computationally equivalent to `[['X', 'XYY', 1]]` in terms of this PS-L-system.

However, there is no restriction that the same instruction cannot be rewritten into more than one forms. Consider the following replacement rule set – [['X', 'XYY'], [['X', 'YYX']]. Given that both rules are of the same priority, the first rule (['X', 'XYY']) automatically takes precedence over the second rule (['X', 'YYX']), implying that the order of such rules is significant (non-communicative). In reality, the second rule (['X', 'YYX']) will never be used.

In terms of implementation, all replacement rules are converted into a 4-element list with the 4th element being 'replacement'. This is to cater for stochastic or probability rules, where the 4th element is 'probability'; and function rules, where the 4th element is 'function'. Hence, there three ways to write the same replacement rule, ['X', 'YYX']:

- [<instruction>, <replaced instruction(s)>]; for example, ['X', 'YYX']
- [<instruction>, <replaced instruction(s)>, <priority>]; for example, ['X', 'YYX', 1]
- [<instruction>, <replaced instruction(s)>, <priority>, 'replacement']; for example, ['X', 'YYX', 1, 'replacement']

The second type of rule is stochastic or probability rule. Each probability rule is defined as a 5-element list in the format of [<instruction>, <replaced instruction(s)>, <priority>, 'probability', <probability>], where <probability> refers to the probability of execution. For example, the rule "for 50% of the time, substitute or rewrite X into XYY" will be implemented as ['X', 'XYY', 1, 'probability', 0.5]. Hence, a replacement rule, such as ['X', 'YYX', 1, 'replacement'], is computationally equivalent to a probability rule of 100%, such as ['X', 'YYX', 1, 'probability', 1.0], and vise versa.

The third type of rule is function rule, which caters for other more complex operations not addressed by replacement or probability rules. These cases include parametric rules, and probabilistic parametric rules. Each function rule is defined as a 4-element list in the format of [<instruction>, <function>, <priority>, 'function'], where <function> is a Python function that takes instructions and position as parameters. For example, the function

```
def replaceFunction(instructions, position):
    if instructions[position+3] == 'O':
        return 'BAAB'
    elif instructions[position-1] == 'O':
        return 'AABB'
    else: return 'OOAB'
```

will perform the following actions, assuming the instruction length to be one and the current instruction is "A":

- replace the current instruction "A" with "BAAB" if the 3rd instruction from the current instruction is "O"; if not,
- replace the current instruction "A" with "AABB" if the preceding instruction from the current instruction is "O"; if not,
- replace the current instruction "A" with "OOAB".

This rule is then implemented as ['A', replaceFunction, 1, 'function'].

The priority of the rules is crucial to the result outcome as shown in Table 1. In event when there is more than one rule with the same priority, the execution order is based on the type of rule – replacement rule to execute first, followed by probability rule, followed by function rule. However, if there is more than one rule type of the same priority; such as, 2 replacement rules of the same priority; the execution order can be indeterministic, resulting in indeterministic results.

The third step is to perform one or more iterations of rewriting, given one or more starting instructions (also known as axiom).

The fourth and last step, which is optional, is to take the iteratively rewritten instruction list and generate a LOGO-like Turtle graphics using Tkinter library within Python Standard Library. The eventual graphics can be saved as a scalable vector graphics (SVG) and the commands used to generate the Turtle graphics can be saved as a Python script file. However, the resulting Python script file can be large as it is a naïve code generation without using loops for code optimization. In order to generate the graphics from the instruction string, a mapping dictionary is required to map each instruction into a Turtle action.

Table 1: Differences in Results from the Same Set of Rules but Different Priorities

Rules	`[['X', 'XYY', 1],` ` ['Y', 'XXY', 2]]`	`[['Y', 'XXY', 1],` ` ['X', 'XYY', 2]]`
Axiom (Generation 0)	X	X
Generation 1	X XXY XXY	XYY
Generation 2	XXX YXX YXX XYX XYX XXY XXY XXY XXX YXX YXX XYX XYX XY	XYY XYY XYY YXY YXY YY
Generation 3	XXX YXX YXX XYX XYX XXY XXY XXY XXX YXX YXX XYX XYX XYX XXY XXY XXX YXX YXX XYX XYX XYX XXY XXY XXX YXX YXX YXX XYX XYX XXY XXY XXX YXX YXX YXX XYX XYX XXY XXY XXY XXX YXX YXX XYX XYX XYX XXY XXY XXX YXX YXX XYX XYX XYX XXY XXY XXX YXX YXX YXX XYX XYX XXY XXY XXX YXX YXX YXX XYX XYX XXY XXY XXY XXX YXX YXX XYX XYX XY	XYY XYY XYY YXY YXY YYX YYX YYX YYY XYY XYY YXY YXY YXY YYX YYX YYY XYY XYY YXY YXY YXY YYX YYX YYY XYY XYY XYY YXY YXY YYX YYX YYY

The following Turtle commands are defined: forward, backward, right (turn), left (turn), pen up, pen down and home. The default mapping is given as:

```
mapping = {'set_angle': 90,
 'random_angle': 0,
 'set_distance': 1,
 'random_distance': 0,
 'set_heading': 0,
 'set_colour': 'black',
 'background_colour': 'ivory',
 'F': 'forward',
 'B': 'backward',
 'R': 'right',
 'L': 'left',
 'H': 'home',
 'U': 'penup',
 'D': 'pendown',
 '[': 'push',
 ']': 'pop'}
```

which can be read as

- a left or right turn is set at 90 degrees (set_angle). This setting is mandatory.
- random angles of turn can be set using 'random_angle', where the actual angle will be from the set_angle to set_angle + random_angle (by uniform distribution). For example, if random_angle is 10 degrees, it means that the actual angle at each turn will be uniformly distributed from 90 to 100 degrees. This will be set to 0 if not given.
- each forward or backward move is set at 1 (set_distance). This setting is mandatory.
- random distance of each move can be set using 'random_distance,' following the same logic as 'random_angle.' This will be set to 0 if not given.
- Turtle is set to head towards 0 degrees (east or to the right of the screen). North (toward top), west (towards left), and south (towards

bottom) are 90, 180, 270 degrees respectively. This will be set to 0 if not given.

- default pen colour can be set using TK colour names as 'set_colour'. This will be set to black if not given. Other colours can be set and any un-used symbols (other than 'F', 'B', 'R', 'L', 'H', 'U', and 'D') can be used to set pen colours (please see http://wiki.tcl.tk/37701 for available colours).
- canvas background colour can be set using background_colour. This will be set to ivory if not given.
- 'F', 'B', 'R', 'L', 'H', 'U', and 'D' represents the Turtle commands of forward, backward, right turn, left turn, home, pen up, and pen down respectively. Home is defined as the start coordinate.
- '[' pushes the current state (position and heading) of the Turtle into the stack.
- ']' pops (in a last in first out manner) and sets the Turtle to the last pushed state without drawing the move.

CODE FILE (LINDENMAYER.PY)

```
'''
Framework for Lindenmayer System (L-System)
Copyright (c) Maurice H.T. Ling <mauriceling@acm.org>
Date created: 4th January 2015
'''
import random

import constants

class lindenmayer(object):
    '''Lindenmayer system, also commonly known as L-
    System,  is developed by Aristid Lindenmayer in 1968
    (reference: Mathematical models for cellular
    interaction in development. Journal of Theoretical
    Biology 18:280-315). It is a set of formal grammar of
    production rules for rewiting an initial axiom or seed
    text over generations.

    This implementation defines 3 types of rules, also
```

known as production rules or predicates; replacement, probability, and function rules. Each rule can be given a priority.

The simplest form of production rule takes the form of 'A -> BAC', which is read as "whenever 'A' is found, it is replaced/rewritten as 'BAC'". For example, if the starting axiom is "A", then the following will happen

- *Generation 0: A*
- *Generation 1: BAC*
- *Generation 2: BBACC*
- *Generation 3: BBBACCC*
- *Generation 4: BBBBACCCC*
- *and so on.*

In this case, the predicate 'A -> BAC' can be written as in 4 different ways - C{['A', 'BAC']}, C{['A', 'BAC', 1]}, C{['A', 'BAC', 1, 'replacement']}, or C{['A', 'BAC', 1, 'replacement', 1]}.

When a list of 2-elements is given (e.g., C{['A', 'BAC']}), it is taken to be replacement rule with the highest priority; that is, priority of 1. Hence, C{['A', 'BAC']}, C{['A', 'BAC', 1]}, C{['A', 'BAC', 1, 'replacement']}, and C{['A', 'BAC', 1, 'replacement', 1]} are the same.

This also means that production rules can have different priorities. For example, given C{[['A', 'BAC', 1], ['B', 'BC', 2]]}, rule C{['A', 'BAC', 1]} will be executed before C{['B', 'BC', 2]} in the following manner

- *Generation 0: A*
- *Generation 1: BCAC # A -> BAC, BAC -> BCAC*
- *Generation 2: BCCBCACC*
- *Generation 3: BCCCBCCBCACCC*
- *and so on as all production rules in ascending order of priorities (with '1' being the highest priority) will be executed in sequence on the resulting axiom at that current point in time. As a result, the order of rules within the same priority is significant if they recognize the same instruction or command.*

However, if given C{[['A', 'BAC', 1], ['B', 'BC',

1]]}, then

- Generation 0: A
- Generation 1: BAC
- Generation 2: BCBACC
- Generation 3: BCCBCBACCC
- Generation 4: BCCCBCCBCBACCCC
- and so on.

The second form of production rule is probabilistic, also known as stochastic grammars. Probabilistic rule will take the format of C{[<domain>, <range>, <priority>, 'probability', <probability>]}. For example, C{['A', 'BAC', 1, 'probability', 0.5]} means that 'A' will only be rewritten into 'BAC' 50% of the time. 'A' will be left unchanged 50% of the time. The same priority principle applies. Hence, C{['A', 'BAC', 1, 'probability', 1]} is in effect the same as C{['A', 'BAC', 1, 'replacement']}.

The third form of production rule is function rule, which takes the form of C{[<domain>, <function>, <priority>, 'function']}. For example, C{['A', axiom_func, 1, 'function']} means that when 'A' is encounted in the axiom, the command string up to that point in time will be used as parameter for axion_func function, such as

- Generation 0: A
- Generation 1: dependent on the return value of axiom_func('A') and so on.

For example, given an axiom of 'ACCCABABDD', and C{['AB', replaceFunction, 1, 'function']} as production rule where replaceFunction is defined as

```
>>> def replaceFunction(instructions, position):
>>>     if instructions[position+3] == 'O': return '
            BAAB'
>>>     elif instructions[position-1] == 'O': return
            'AABB'
>>>     else: return 'OOAB'
```

- Generation 0: ACCCABABDD
- Generation 1: ACCCOOABOOABDD
- Generation 2: ACCCOOBAABOOAABBDD
- Generation 3: ACCCOOBABAABOOAABBDD
- Generation 4: ACCCOOBABABAABOOAABBDD

```
        - Generation 5: ACCCOOBABABABAABOOAABBDD

    In summary, the following rule formats are allowed:
        - C{[<domain>, <range>]}
        - C{[<domain>, <range>, <priority>]}
        - C{[<domain>, <range>, <priority>,
        'replacement']}
        - C{[<domain>, <range>, <priority>, 'probability',
        <probability>]}
        - C{[<domain>, <function>, <priority>,
        'function']}
        '''
    def __init__(self, command_length=1):
        '''Constructor method.

        @param command_length: length of each instruction
        or command. Default = 1
        @type command_length: integer'''
        self.command_length = command_length
        self.rules = []

    def add_rules(self, rules):
        '''Method to add a list of production rules /
        predicates into the system.

        @param rules: a list of list describing the
        production rules. Please see above for rule
        syntax.
        @type rules: list'''
        for x in rules:
            if len(x) == 2:
                # [predicate, replacement]
                self.rules.append([x[0], x[1], 1,
                                   'replacement', None])
            elif len(x) == 3:
                # [predicate, replacement, priority]
                self.rules.append([x[0], x[1], int(x[2]),
                                   'replacement', None])
            elif len(x) == 4 and (x[3] not in
                ['probability', 'replacement',
                 'function']):
                # [predicate, replacement, priority,
                # '<something_else>']
                print('''Warning: Rule type can only be
'probabilistic', 'replacement' or 'function'. Rule, %s, is
not added into system.''' % str(x))
            elif len(x) == 4 and x[3] == 'replacement':
                # [predicate, replacement, priority,
```

```
            # 'replacement']
            self.rules.append([x[0], x[1], int(x[2]),
                               x[3], None])
        elif len(x) == 4 and x[3] == 'probability':
            # [predicate, replacement, priority,
            # 'probability']
            print('''Warning: Function rule will
require a probability. Rule, %s, is added into system as a
replacement rule (100% activation probability).''' %
str(x))
            self.rules.append([x[0], x[1], int(x[2]),
                               'replacement', None])
        elif len(x) == 5 and x[3] == 'probability':
            # [predicate, replacement, priority,
            # 'probability', probability]
            self.rules.append([x[0], x[1], int(x[2]),
                               x[3], float(x[4])])
        elif len(x) == 4 and x[3] == 'function':
            # [predicate, function, priority,
            # 'function']
            self.rules.append([x[0], x[1], int(x[2]),
                               x[3], None])
    self.priority_levels = [x[2]
                            for x in self.rules][-1]

def _apply_priority_rules(self, priority,
                          data_string):
    '''Private method - to be used by apply_rules
    method to apply production rules of a particular
    priority.

    @param priority: order of priority
    @type priority: integer
    @param data_string: data or symbol string to be
    processed
    @type data_string: string
    @return: rewritten data_string
    '''
    rules = [x for x in self.rules
             if x[2] == int(priority)]
    ndata = ''
    pointer = 0
    while pointer < len(data_string):
        cmd = data_string[pointer:
                          pointer+self.command_length]
        for rule in rules:
            if cmd == rule[0] and \
            rule[3] == 'replacement':
```

```
                    cmd = rule[1]
                    break
                if cmd == rule[0] and \
                rule[3] == 'probability' \
                and random.random() < x[4]:
                    cmd = rule[1]
                    break
                if cmd == rule[0] and \
                   rule[3] == 'function':
                    cmd = rule[1](data_string, pointer)
                    break
            if cmd == None: cmd = ''
            ndata = ndata + cmd
            pointer = pointer + self.command_length
        return ndata

    def _apply_rules(self, data_string):
        '''Private method - to apply all production rules
        on axiom string (in the first generation) or
        data/symbol string (in the subsequent
        generations).
        This method is implemented as a generator.

        @param data_string: data or symbol string to be
        processed
        @type data_string: string
        @return: rewritten data_string'''
        for priority in
        list(range(1, self.priority_levels+1)):
            data_string = \
                self._apply_priority_rules(priority,
                                           data_string)
        return data_string

    def generate(self, axiom, iterations):
        '''Method to apply all production rules on an
        initial axiom string over a number of iterations.

        @param axiom: data or symbol string to be
        processed
        @type axiom: string
        @param iterations: number of repetitions /
        iterations
        @type iterations: integer
        @return: rewritten axiom'''
        self.axiom = axiom
        iterations = int(iterations)
        count = 1
```

```
    while count < iterations + 1:
        self.axiom = self._apply_rules(self.axiom)
        print('Generation %s: Axiom length = %s' % \
              (str(count), str(len(self.axiom))))
        count = count + 1
    return self.axiom

def turtle_generate(self, scriptfile=None,
                    imagefile=None, start=(0, 0),
                    mapping={}, data_string=None):
    '''Method for naive code generation to visualize
    the data or symbol string using Turtle graphics.
    This method generates the Python codes for Turtle
    graphics using the TK Turtle graphics module, and
    prints out the resulting Python code as a file.

    This method does not use any loops to reduce
    repetitive Turtle commands; hence, the resulting
    code file can be huge.

    A mapping dictionary is used to convert the symbol
    string into Turtle commands. The following Turtle
    commands are defined: forward, backward, right
    (turn), left (turn), pen up, pen down and home.
    The
    default mapping is given as

    >>> mapping = {'set_angle': 90,
    >>>            'random_angle': 0,
    >>>            'set_distance': 1,
    >>>            'random_distance': 0,
    >>>            'set_heading': 0,
    >>>            'set_colour': 'black',
    >>>            'background_colour': 'ivory',
    >>>            'F': 'forward',
    >>>            'B': 'backward',
    >>>            'R': 'right',
    >>>            'L': 'left',
    >>>            'H': 'home',
    >>>            'U': 'penup',
    >>>            'D': 'pendown',
    >>>            '[': 'push',
    >>>            ']': 'pop'}

    which can be read as
        - a left or right turn is set at 90 degrees
        (set_angle). B{This setting is mandatory.}
        - random angles of turn can be set using
```

```
        'random_angle', where the actual angle will be
        from the set_angle to set_angle + random_angle
        (by uniform distribution). For example, if
        random_angle is 10 degrees, it means that the
        actual angle at each turn will be uniformly
        distributed from 90 to 100 degrees. B{This
        will be set to 0 if not given.}
        - each forward or backward move is set at 1
        (set_distance). B{This setting is mandatory.}
        - random distance of each move can be set
        using 'random_distance', following the same
        logic as 'random_angle'. B{This will be set to
        0 if not given.}
        - turtle is set to head towards 0 degrees
        (east or to the right of the screen). North
        (toward top), west (towards left), and south
        (towards bottom) are 90, 180, 270 degrees
        respectively. B{This will be set to 0 if not
        given.}
        - default pen colour can be set using TK
        colour names as 'set_colour'. B{This will be
        set to black if not given.} Other colours can
        be set and any un-used symbols (other than
        'F', 'B', 'R', 'L', 'H', 'U', and 'D') can be
        used to set pen colours (please see
        http://wiki.tcl.tk/37701 for available
        colours).
        - canvas background colour can be set using
        background_colour. B{This will be set to ivory
        if not given.}
        - 'F', 'B', 'R', 'L', 'H', 'U', and 'D'
        represents the Turtle commands of forward,
        backward, right turn, left turn, home, pen up,
        and pen down respectively. Home is defined as
        the start coordinate.
        - '[' pushes the current state (position and
        heading) of the Turtle into the stack.
        - ']' pops (in a last in first out manner) and
        sets the Turtle to the last pushed state
        without drawing the move.

    @param scriptfile: file name to write out the
    Turtle commands. Default = None, no file will be
    written
    @type scriptfile: string
    @param imagefile: SVG file name to write out
    Turtle graphics. Default = None, no file will be
    written
```

```
@type imagefile: string
@param start: starting or home coordinate. Default
= (0, 0) which is the centre of the TK window
@type start: tuple
@param mapping: map to convert the symbol string
into Turtle commands. Please see explanation
above.
@type mapping: dictionary
@param data_string: data or symbol string to be
processed. Default = None, internally stored axiom
string (by lindenmayer.generate() method) will be
used instead.
@type data_string: string
@return: Python script file of Turtle commands'''
if len(mapping) == 0:
    mapping = {'set_angle': 90,
               'random_angle': 0,
               'set_distance': 1,
               'random_distance': 0,
               'set_heading': 0,
               'set_colour': 'black',
               'background_colour': 'ivory',
               'F': 'forward',
               'B': 'backward',
               'R': 'right',
               'L': 'left',
               'H': 'home',
               '[': 'push',
               ']': 'pop'}
if data_string == None:
    data_string = self.axiom
stack = []
if 'random_angle' not in mapping:
    mapping['random_angle'] = 0
if 'random_distance' not in mapping:
    mapping['random_distance'] = 0
if 'set_heading' not in mapping:
    mapping['set_heading'] = 0
if 'set_colour' not in mapping:
    mapping['set_colour'] = 'black'
if 'background_colour' not in mapping:
    mapping['background_colour'] = 'ivory'
data_string = [cmd for cmd in data_string
               if cmd in mapping]
if scriptfile != None:
    f = open(scriptfile, 'w')
    f.write("''' \n")
    f.write('Turtle Graphics Generation from \
```

```
                Lindenmayer System \n')
        f.write('in COPADS (http://github.com/ \
                copads/copads) \n\n')
        f.write('Code length = %s \n' % \
                str(len(data_string)))
        f.write('Code string = %s \n' % \
                ''.join(data_string))
        f.write('Code mapping = %s \n' % (mapping))
        f.write("''' \n\n")
        f.write('import turtle \n\n')
        f.write('t = turtle.Turtle() \n')
        f.write('t.setundobuffer(1) \n')
        f.write("turtle.bgcolor('%s') \n" % \
                mapping['background_colour'])
        f.write('t.speed(0) \n\n')
        f.write('t.setheading(%s) \n' % \
                float(mapping['set_heading']))
        f.write('t.penup() \n')
        f.write('t.setposition(%s, %s) \n' % start)
        f.write("t.pencolor('%s') \n" % \
                mapping['set_colour'])
        f.write('t.pendown() \n\n')
    import turtle
    t = turtle.Turtle()
    t.setundobuffer(1)
    exec("turtle.bgcolor('%s')" % \
         mapping['background_colour'])
    exec('t.speed(0)')
    exec('t.setheading(%s)' % \
         float(mapping['set_heading']))
    exec('t.penup()')
    exec('t.setposition(%s, %s)' % start)
    exec("t.pencolor('%s')" % mapping['set_colour'])
    exec('t.pendown()')
    count = 0
    for cmd in data_string:
        count = count + 1
        if count % 1000 == 0:
            print('%s instructions processed ...' % \
                  str(count))
        if mapping[cmd] == 'push':
            status = [t.position(), t.heading()]
            stack.append(status)
        if mapping[cmd] == 'pop':
            try:
                status = stack.pop()
                if scriptfile != None:
                    f.write('t.penup() \n')
```

```
                    f.write('t.setposition(%s, %s) \n'
\
                            % status[0])
                    f.write('t.setheading(%s) \n' % \
                            status[1])
                    f.write('t.pendown() \n')
                exec('t.penup()')
                exec('t.setposition(%s, %s)' % \
                     status[0])
                exec('t.setheading(%s)' % status[1])
                exec('t.pendown()')
            except IndexError: pass
        if mapping[cmd] == 'forward':
            distance = mapping['set_distance'] + \
            random.random()*mapping['random_distance']
            if scriptfile != None:
                f.write('t.forward(%s) \n' % \
                        (distance))
            exec('t.forward(%s)' % (distance))
        if mapping[cmd] == 'backward':
            distance = mapping['set_distance'] + \
            random.random()*mapping['random_distance']
            if scriptfile != None:
                f.write('t.backward(%s) \n' % \
                        (distance))
            exec('t.backward(%s)' % (distance))
        if mapping[cmd] == 'home':
            if scriptfile != None:
                f.write('t.penup() \n')
                f.write('t.setposition(%s, %s) \n' % \
                        start)
                f.write('t.pendown()')
            exec('t.penup()')
            exec('t.setposition(%s, %s)' % start)
            exec('t.pendown() \n')
        if mapping[cmd] == 'right':
            angle = mapping['set_angle'] + \
              random.random()*mapping['random_angle']
            if scriptfile != None:
                f.write('t.right(%s) \n' % str(angle))
            exec('t.right(%s)' % str(angle))
        if mapping[cmd] == 'left':
            angle = mapping['set_angle'] + \
              random.random()*mapping['random_angle']
            if scriptfile != None:
                f.write('t.left(%s) \n' % str(angle))
            exec('t.left(%s)' % str(angle))
        if mapping[cmd] in constants.TKColours:
```

```
                f.write("t.pencolor('%s') \n" % \
                        mapping[cmd])
                exec("t.pencolor('%s')" % mapping[cmd])
        exec('t.penup()')
        exec('t.hideturtle()')
        if imagefile != None:
            exec('import canvasvg')
            exec("canvasvg.saveall('%s',
                 t.getscreen()._canvas)" % imagefile)
            f.write('\n')
            f.write('try: \n')
            f.write('    import canvasvg \n')
            f.write("    canvasvg.saveall('%s',
                         t.getscreen()._canvas) \n" \
                    % imagefile)
            f.write('except ImportError: pass \n')
        exec('turtle.done()')
        print('%s instructions processed. Drawing
              completed.' % str(count))
        if scriptfile != None:
            f.write('\n')
            f.write('t.penup() \n')
            f.write('t.hideturtle() \n')
            f.write('turtle.done() \n')
            f.close()
```

TEST CASES (T-LINDENMAYER.PY)

```
import sys
import os
import unittest
import re

sys.path.append(os.path.join(os.path.dirname(os.getcwd()),
'copads'))
import lindenmayer as N

class testReplacement(unittest.TestCase):
    '''
    Test for single replacement rule without priority.
    Command length = 1
    '''
    def setUp(self):
        s = N.lindenmayer(1)
        r = [['A', 'BAC']]
        s.add_rules(r)
```

```
        axiom = 'A'
        self.result = []
        for i in range(10):
            axiom = s._apply_rules(axiom)
            self.result.append(axiom)
        self.answer = ['BAC',
                       'BBACC',
                       'BBBACCC',
                       'BBBBACCCC',
                       'BBBBBACCCCC',
                       'BBBBBBACCCCCC',
                       'BBBBBBBACCCCCCC',
                       'BBBBBBBBACCCCCCCC',
                       'BBBBBBBBBACCCCCCCCC',
                       'BBBBBBBBBBACCCCCCCCCC']
    def testGeneration(self):
        self.assertEqual(self.result, self.answer)

class testReplacementPriority(unittest.TestCase):
    '''
    Test for replacement rules with priority.
    Command length = 1
    '''
    def setUp(self):
        s = N.lindenmayer(1)
        r = [['A', 'BAC', 1],
             ['B', 'BC', 2]]
        s.add_rules(r)
        axiom = 'A'
        self.result = []
        for i in range(8):
            axiom = s._apply_rules(axiom)
            self.result.append(axiom)
        self.answer = ['BCAC',
                       'BCCBCACC',
                       'BCCCBCCBCACCC',
                       'BCCCCBCCCBCCBCACCCC',
                       'BCCCCCBCCCCBCCCBCCBCACCCCC',
'BCCCCCCBCCCCCBCCCCBCCCBCCBCACCCCCC',
'BCCCCCCCBCCCCCCBCCCCCBCCCCBCCCBCCBCACCCCCCC',
'BCCCCCCCCBCCCCCCCBCCCCCCBCCCCCBCCCCBCCCBCCBCACCCCCCCC']
    def testGeneration(self):
        self.assertEqual(self.result, self.answer)

class testReplacementNoPriority(unittest.TestCase):
    '''
    Test for replacement rules without priority.
    Command length = 1
```

```
    '''
    def setUp(self):
        s = N.lindenmayer(1)
        r = [['A', 'BAC'],
             ['B', 'BC']]
        s.add_rules(r)
        axiom = 'A'
        self.result = []
        for i in range(8):
            axiom = s._apply_rules(axiom)
            self.result.append(axiom)
        self.answer = ['BAC',
                       'BCBACC',
                       'BCCBCBACCC',
                       'BCCCBCCBCBACCCC',
                       'BCCCCBCCCBCCBCBACCCCC',
                       'BCCCCCBCCCCBCCCBCCBCBACCCCCC',
               'BCCCCCCBCCCCCBCCCCBCCCBCCBCBACCCCCCC',
     'BCCCCCCCBCCCCCCBCCCCCBCCCCBCCCBCCBCBACCCCCCCC']
    def testGeneration(self):
        self.assertEqual(self.result, self.answer)

class testReplacement2(unittest.TestCase):
    '''
    Test for replacement rules without priority.
    Command length = 2
    '''
    def setUp(self):
        s = N.lindenmayer(2)
        r = [['AA', 'BBAAC'],
             ['AB', 'ABC'],
             ['AC', 'BCC']]
        s.add_rules(r)
        axiom = 'AA'
        self.result = []
        for i in range(8):
            axiom = s._apply_rules(axiom)
            self.result.append(axiom)
        self.answer = ['BBAAC',
                       'BBBBAACC',
                       'BBBBBBAACCC',
                       'BBBBBBBBAACCCC',
                       'BBBBBBBBBBAACCCCC',
                       'BBBBBBBBBBBBAACCCCCC',
                       'BBBBBBBBBBBBBBAACCCCCCC',
                       'BBBBBBBBBBBBBBBBAACCCCCCCC']
    def testGeneration(self):
        self.assertEqual(self.result, self.answer)
```

```
class testFunction(unittest.TestCase):
    '''
    Test for function rules.
    Command length = 1
    '''
    def replaceFunction(self, dstring, position):
        if dstring[position+3] == 'O': return 'BAAB'
        elif dstring[position-1] == 'O': return 'AABB'
        else: return 'OOAB'
    def setUp(self):
        s = N.lindenmayer(2)
        r = [['AB', self.replaceFunction, 1, 'function']]
        s.add_rules(r)
        axiom = 'ACCCABABDD'
        self.result = []
        for i in range(5):
            axiom = s._apply_rules(axiom)
            self.result.append(axiom)
        self.answer = ['ACCCOOABOOABDD',
                       'ACCCOOBAABOOAABBDD',
                       'ACCCOOBABAABOOAABBDD',
                       'ACCCOOBABABAABOOAABBDD',
                       'ACCCOOBABABABAABOOAABBDD']
    def testGeneration(self):
        self.assertEqual(self.result, self.answer)


if __name__ == '__main__':
    unittest.main()
```

DEMONSTRATION

Two demonstrations are performed using Sierpinski Triangle and Tree. Sierpinski Triangle, also known as Sierpinski Sieve, is a symmetrical fractal design. Structural symmetry is maintained by minimizing the error in each angle. In this demonstration, error is gradually increased from no error, where all angles are 60 degrees; to 12.5% random error, which translates to 60 degrees ± 7.5 degrees. The code for Sierpinski Triangle without randomness in angles is shown as

```
import random
from lindenmayer import lindenmayer
```

```
axiom = 'A'
rules = [['A', 'BLALB'],
 ['B', 'ARBRA']]
start_position = (-300, 250)
iterations = 8
turtle_file = '19_lindenmayer_sierpinski_turtle.py'
image_file = '19_lindenmayer_sierpinski_turtle.svg'
mapping = {'set_angle': 60,
 'random_angle': 0,
 'set_distance': 2.5,
 'random_distance': 0,
 'background_colour': 'RoyalBlue1',
 'A': 'forward',
 'B': 'forward',
 'R': 'right',
 'L': 'left'}

l = lindenmayer(1)
l.add_rules(rules)
l.generate(axiom, iterations)
l.turtle_generate(turtle_file, image_file, start_position,
mapping)
```

The generated Turtle graphics (Figure 1) demonstrated that the effects of random angles on the appearance of Sierpinski Triangle. Minor overlaps can be seen even at a 2.5% randomness. By 7.5% randomness, the geometry appears to be collapsing on itself. By 12.5%, it is almost unrecognizable as a Sierpinski Triangle. On the other hand, randomness in angles does not appear to affect the "algorithmic beauty" of the tree. The code for the tree without random angle is

```
import random
from lindenmayer import lindenmayer

axiom = 'F'
rules = [['F', '0FFL[1LFRFRF]R[2RFLFLF]']]
start_position = (0, -200)
iterations = 5
turtle_file = '19_lindenmayer_tree_turtle.py'
image_file = '19_lindenmayer_tree_turtle.svg'
mapping = {'set_angle': 22,
 'random_angle': 0,
 'set_distance': 5,
 'random_distance': 0,
```

```
 'set_heading': 90,
 'background_colour': 'ivory',
 'F': 'forward',
 'R': 'right',
 'L': 'left',
 '[': 'push',
 ']': 'pop',
 '0': 'brown',
 '1': 'dark green',
 '2': 'forest green'}

l = lindenmayer(1)
l.add_rules(rules)
l.generate(axiom, iterations)
l.turtle_generate(turtle_file, image_file, start_position,
mapping)
```

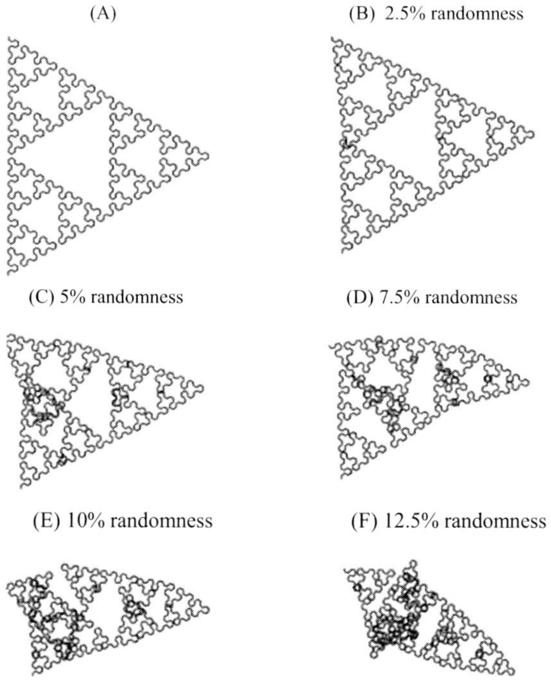

Figure 1. Sierpinski triangle with varying degrees of angle randomness. Each of these images are generated with 6 generations at 60° angle, with or without randomness in angles. Panel A shows no randomness and all angles are 60° angle. Panel B shows 2.5% randomness in angles (60° ± 1.5°). Panels C, D, E, and F show 5% (60° ± 3.0°), 7.5% (60° ± 4.5°), 10% (60° ± 6.0°), and 12.5% (60° ± 7.5°) randomness in angles respectively.

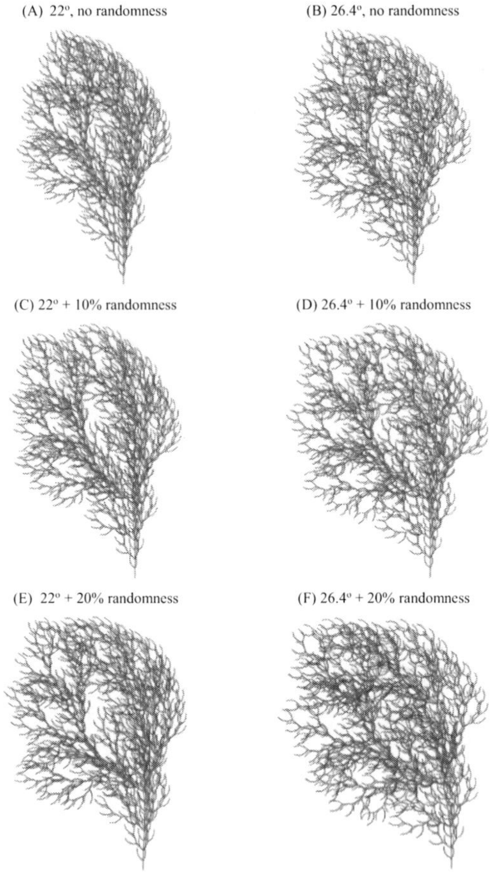

Figure 2. Tree with varying degrees of angles and randomness. Each image is generated 4 generations. Panels A and B show 22°, and 26.4° (20% increase of 20% increase in angle from 22°) respectively. Panels C and E show 22° angle with 10% (22° ± 2.2°), and 20% (22° ± 4.4°) randomness respectively. Panels D and F show 26.4° angle with 10% (26.4° ± 2.64°), and 20% (26.4° ± 5.28°) randomness respectively.

The generated Turtle graphics for the tree (Figure 2) shows that randomness in angles does not affect the "feel" and beauty of the tree even at 20% random angles, which is remarkably different from Sierpinski Triangle (unrecognizable at 12.5% randomness in angles). This suggests that algorithmic beauty can exist in two different ways – symmetric beauty as in the case of Sierpinski Triangle and overall structural beauty as in the case of the tree.

REFERENCES

Boudon, F, Pradal, C, Cokelaer, T, Prusinkiewicz, P, and Godin, C. (2012) L-Py: An L-System Simulation Framework for Modeling Plant Architecture Development Based on a Dynamic Language, *Frontiers in Plant Science* 3, 76.

Cici, SZH., Adkins, S, and Hanan, J. (2009) Modelling the morphogenesis of annual sowthistle, a common weed in crops, *Computers and Electronics in Agriculture* 69, 40-45.

Davoodi, A, and Boozarjomehry, RB. (2016). Developmental model of an automatic production of the human bronchial tree based on L-system. *Computer Methods and Programs in Biomedicine* 132, 1–10.

Eichhorst, P, and Ruskey, F. (1981) On unary stochastic Lindenmayer systems, *Information and Control* 48, 1-10.

Fridenfalk, M. (2016). Application for Real-Time Generation of Mathematically Defined 3D Worlds. In *Transactions on Computational Science 28: Special Issue on Cyberworlds and Cybersecurity*, M. L. Gavrilova, C. J. K. Tan, and A. Sourin, eds. (Berlin, Heidelberg: Springer Berlin Heidelberg), pp. 45–68.

Galarreta-Valverde, MA, Macedo, MM, Mekkaoui, C, and Jackowski, MP. (2013) Three-dimensional synthetic blood vessel generation using stochastic L-systems, In *SPIE Medical Imaging*, pp 86691I-86691I-86696, International Society for Optics and Photonics.

Hoyal Cuthill, JF, and Conway Morris, S. (2014) Fractal branching organizations of Ediacaran rangeomorph fronds reveal a lost Proterozoic body plan, *Proceedings of the National Academy of Sciences of the United States of America* 111, 13122-13126.

Lindenmayer, A. (1968a) Mathematical models for cellular interactions in development. I. Filaments with one-sided inputs, *Journal of Theoretical Biology* 18, 280-299.

Lindenmayer, A. (1968b) Mathematical models for cellular interactions in development. II. Simple and branching filaments with two-sided inputs, *Journal of Theoretical Biology* 18, 300-315.

Lindenmayer, A. (1971) Developmental systems without cellular interactions, their languages and grammars, *Journal of Theoretical Biology* 30, 455-484.

Lindenmayer, A. (1975) Developmental algorithms for multicellular organisms: a survey of L-systems, *Journal of Theoretical Biology* 54, 3-22.

Room, P, Hanan, J, and Prusinkiewicz, P. (1996) Virtual plants: new perspectives for ecologists, pathologists and agricultural scientists, *Trends in Plant Science* 1, 33-38.

Parish, YI, and Müller, P. (2001) Procedural modeling of cities, In *Proceedings of the 28th Annual Conference on Computer Graphics and Interactive Techniques,* pp 301-308, ACM.

Perttunen, J, and Sievänen, R. (2005) Incorporating Lindenmayer systems for architectural development in a functional-structural tree model, *Ecological Modelling* 181, 479-491.

Prince, DR, Fletcher, ME, Shen, C, and Fletcher, TH. (2014) Application of L-systems to geometrical construction of chamise and juniper shrubs, *Ecological Modelling* 273, 86-95.

Prusinkiewicz, P, and Lindenmayer, A. (1990) *The algorithmic beauty of plants,* Springer-Verlag.

Rongier, G, Collon, P, and Renard, P. (2017a). Stochastic simulation of channelized sedimentary bodies using a constrained L-system. *Computers & Geosciences* 105, 158–168.

Rongier, G, Collon, P, and Renard, P. (2017b). A geostatistical approach to the simulation of stacked channels. *Marine and Petroleum Geology* 82, 318–335.

Samal, A, Peterson, B, and Holliday, DJ. (1994) Recognizing plants using stochastic L-systems, In. *Proceedings of IEEE International Conference on Image Processing*, 1994 (ICIP-94), pp 183-187, IEEE.

Shi, Y, Cheng, X, and Zhang, H. (2011) Three dimensional trees emulation based on Parametric L-system, *Journal of Tongji University.* Natural Science 39, 1871-1876.

Taralova, EH, Schlecht, J, Barnard, K, and Pryor, BM. (2011) Modelling and visualizing morphology in the fungus Alternaria, *Fungal Biology* 115, 1163-1173.

van Dalen, D. (1971) A note on some systems of Lindenmayer, *Theory of Computing Systems* 5, 128-140.

Xin, L, Xu, L, Li, D, and Fu, D. (2014) The 3D reconstruction of greenhouse tomato plant based on real organ samples and parametric L-system, In *Sixth International Conference on Digital Image Processing, International Society for Optics and Photonics*.

Date Submitted: September 13, 2017. *Date Accepted:* December 31, 2017.

In: Current STEM. Volume 1
Editor: Maurice HT Ling
ISBN: 978-1-53613-416-2

Chapter 5

COPADS VI: Fixed Time-Step ODE Solvers with Mixed ODE and Non-ODE Function, and Script Generator

Maurice HT Ling*
Colossus Technologies LLP, Singapore
School of BioSciences, The University of Melbourne, Australia

Abstract

Ordinary differential equations (ODEs) are commonly used in mathematical modelling. However, the standard means of implementing a system of ODEs in Python, using Scipy, does not allow for each ODE to be implemented as an individual Python function. This results in poor documentation and maintainability. We have re-implemented 11 fixed-steps ODE solvers to allow for an ODE system to be implemented as a set of Python functions. Here, the ODE solvers are enhanced to take on a non-ODE function, allowing for modification of the ODE result vector at each time step, which may be useful in cases where one or more results has to be calibrated using other results. In addition, a script generator is implemented to assist in the generation of a Python ODE script from a set

* Corresponding Author Emails: mauriceling@acm.org, mauriceling@colossus-tech.com.

of parameters. This module had been incorporated into the Collection of Python Algorithms and Data Structures (COPADS; https://github.com / mauriceling/copads), under Python Software Foundation License version 2.

Keywords: fixed time step, ODE solver, modifying function

INTRODUCTION

Mathematical modelling is an important aspect of science (Otto and Day, 2007) and differential equation, especially ordinarily differential equation (ODE), is one of the two main paradigms in mathematical modelling (Ling, 2016a); the other being state transition modelling (Chay et al., 2016). As previously described (Ling, 2016b), the standard way of writing ODE systems in Python is via the use of SciPy (Oliphant, 2007) or Odespy (Langtangen and Wang, 2014) to implement one or more ODEs within a single function. Using a simple 3-equation ODE to model zombie invasion (Munz et al., 2009) as an example, the following are means of writing the ODEs in SciPy:

```
def example1(y, t):
    human = birth - transmission*y[0]*y[1] - death*y[0]
     zombie = transmission*y[0]*y[1] + resurect*y[2] - \
               destroy*y[0]*y[1]
     dead = death*y[0] + destroy*y[0]*y[1] - resurect*y[2]
    return [human, zombie, dead]

def example2(y, t):
    f = [birth - transmission*y[0]*y[1] - death*y[0],
         transmission*y[0]*y[1] + resurect*y[2] - \
             destroy*y[0]*y[1],
         death*y[0] + destroy*y[0]*y[1] - resurect*y[2]]
    return f
```

However, standard means of writing ODEs in SciPy (Oliphant, 2007) as shown above present two issues. Firstly, it is common to have an ODE system comprising of many individual ODEs. This will result in function codes spanning multiple lines and possibly across multiple pages, which is difficult to read. Secondly, it does not allow for effective documentation – both at the equation level and the term (sub-equation) level. As a result, a large system of ODEs written in SciPy's method will be difficult to read and usually suffer from the lack of ample documentation.

The ideal solution is to allow users to implement each ODE as a separate function. Besides being able to separate each term within an ODE into meaningful variable names (resulting in self-documenting code), this also allows for function-level documentation that is compatible to documentation generators, such as Epydoc (http://epydoc.sf.net) or Sphinx (http://sphinx-doc.org). This is then followed by consolidating individual ODEs into a system of ODEs using a data structure, such as a list, before solving. In this case, the 3 ODEs in Munz et al. (2009) can be written as:

```
def human(t, y):
    '''Modeling the number of remaining humans.
    Infected represents the number of humans infected
    by existing zombies (zombified). Dead represents
    natural (non-zombified) deaths.'''
    infected = transmission * y[0] * y[1]
    dead = death * y[0]
    return birth - infected - dead
def zombie(t, y):
    '''Modeling the number of zombies. Newly infected
    represents the number of new additions as humans
    are infected by existing zombies (zombified).
    Resurrected represents revival of the dead.
    Destroyed represents the number of zombies
    destroyed by humans.'''
    newly_infected = transmission * y[0] *  y[1]
    resurrected = resurrect * y[2]
    destroyed = destroy * y[0] * y[1]
    return newly_infected + resurrected - destroyed
def dead(t, y):
    '''Modeling the total number of dead. Natural
    death represents natural (non-zombified) deaths
    of humans. Destroyed zombies represent the number
    of zombies destroyed by humans. Created zombies
    represents revival of all dead, be it from natural
    death or destroyed zombies.'''
    natural_death = death * y[0]
    destroyed_zombies = destroy * y[0] * y[1]
    created_zombies = resurrect * y[2]
```

```
    return natural_death + destroyed_zombies - \
           created_zombies
example4 = [human, zombie, dead]
```

However, this will not execute in SciPy. A compromise solution may be using functions within a function as shown,

```
def example3(y, t):
    def human(y, t):
        infected = transmission * y[0] * y[1]
        dead = death * y[0]
        return birth - infected - dead
    def zombie(y, t):
        newly_infected = transmission * y[0] * y[1]
        resurrected = resurrect * y[2]
        destroyed = destroy * y[0] * y[1]
        return newly_infected + resurrected - \
               destroyed
    def dead(y, t):
        natural_death = death * y[0]
        destroyed_zombies = destroy * y[0] * y[1]
        created_zombies = resurrect * y[2]
        return natural_death + destroyed_zombies - \
               created_zombies
    return [human, zombie, dead]
```

However, this results in type error in SciPy; hence, not suitable. Hence, a set of ODE solvers, which allows for each ODE to be implemented as a separate Python function (the ideal solution), has been implemented (Ling, 2016b).

This article documents the code implementation of the ODE solvers in Ling (2016b). At the same time, the ODE solvers are enhanced to take on a non-ODE function (see Appendix A for example code). This allows for modification of the ODE result vector at each time step, which may be useful in cases where one or more results has to be calibrated using other results. In addition, a script generator is implemented to assist in the generation of a Python ODE script from a set of parameters. This module had been incorporated into the Collection of Python Algorithms and Data Structures (COPADS; https://github.com/mauriceling/copads), under Python Software Foundation License version 2.

CODE DESCRIPTION

Eleven fixed time-step ODE solvers were implemented has been implemented in Ling (2016b); namely,

1. Euler method,
2. Heun's method,
3. 3rd order Runge-Kutta method (RK3),
4. 4th order Runge-Kutta method (RK4),
5. 4th order Runge-Kutta method with 3/8 rule (RK4 (3/8)),
6. 4th Runge-Kutta-Fehlberg method (RKF4),
7. 5th order Runge-Kutta-Fehlberg method (RKF5),
8. 4th order Cash-Karp method (Cash and Karp, 1990) (CK4),
9. 5th order Cash-Karp method (Cash and Karp, 1990) (CK5),
10. 4th order Dormand-Prince method (Dormand and Prince, 1980) (DP4), and
11. 5th order Dormand-Prince method (Dormand and Prince, 1980) (DP5).

Each ODE solver implemented in Ling (2016b) consists of a ODE solver core function (known as `solver` function) and a driver, where the ODE solver core function is implemented as a sub-function within the driver function. The ODE solver core function is implemented from the respective Butcher Tableaues (Schober et al., 2014) of the ODE solvers. On the other hand, the driver function of each solver (named as their own solver names, such as `Heun` for Heun's method) is identical for all solvers as the following:

```
driver_function(funcs, x0, y0, step, xmax, nonODEfunc,
                lower_bound, upper_bound,
                overflow, zerodivision):
    yield [x0] + y0
    while x0 < xmax:
        y1 = solver(funcs, x0, y0, step)
        y1 = nonODEfunc(y1, step)
        y1 = boundary_checker(y1, lower_bound, 'lower')
        y1 = boundary_checker(y1, upper_bound, 'upper')
        y0 = y1
```

```
        x0 = x0 + step
        yield [x0] + y0
```

where

- funcs is system of differential equations in a Python list
- x0 is the initial value of x-axis, which is usually starting time
- y0 is a vector of initial float values for variables
- step is the step size on the x-axis (also known as step in calculus)
- xmax is the maximum value of x-axis, which is usually ending time
- nonODEfunc is a non-ODE Python function to modify the variable list (y0)
- lower_bound is set of values, as a Python dictionary, for lower boundary of variables (see Appendix B for example codes)
- upper_bound is set of values, as a Python dictionary for upper boundary of variables (see Appendix B for example codes)
- overflow is a value (usually a large value) to assign in event of over flow error (usually caused by a large number) during integration
- zerodivision is value (usually a large value) to assign in event of zero division error, which results in positive infinity, during integration

In essence, the solver core function performs a single time step solution. The result vector then underwent 3 post-processing steps. Firstly, a user-defined function (known as `nonODEfunc` function; see Appendix A for example code) will be executed. This function will not be part of the ODE system of equations or functions; hence, can be included to modify the result vector. For example, this function can be used to modify a specific value in the result vector with respect to other values in the result vector. Secondly, the values in the result vector are lower bounded as values below the lower bound will be reset to a lower bound value. Lastly, the values in the result vector are upper bounded as values above the upper bound will be reset to an upper bound value. In both cases of the upper and lower bounding, each value in the result vector can be set to a different upper and/or lower bound value and the respective resetting value.

In order for each ODE to be used and substituted in various codes as seamlessly as possible, each driver function uses the same set of parameters. In addition, each driver function is implemented as a generator function to cater for long or virtually infinite simulation, which simulation results can be generated as a data stream for continual processing.

Ling (2016b) also demonstrated that self-documenting ODE solution results can be achieved by non-ODE-bounded variables. That is, one or more parameters be implemented within the variable vector; thereby, resulting in more elements in the variable vector compared to the number of ODE equations (see Appendix C for example code). This will result in the parameters being "printed" along with the simulation results; thus, resulting in self-documentation as the ODE function becomes an operational scaffold. Using non-ODE-bounded variables can be useful in event whereby many sets of simulation results are to be generated, such as in the case of sensitivity analysis (Bao and Zhang, 2014).

Another enhancement to the ODE system is the ODE script generator. A system of ODE can be specified as a Python dictionary where each key-value pair represents an ODE. The key is the name of the ODE function while the value is a Python list where term in the ODE is an element in the list. For example, the following ODE to describe an enzymatic reaction: $A \xrightarrow{enzyme} P$, where molecule A (the reactant) is converted into molecule P (the product) via an enzyme as the catalyst. The formation of molecule P can be modelled as the following ODE,

$$\frac{d\,P}{dt} = (7 * enzyme * A) - (1e^{-6} * P)$$

which can be specified as

```
{'P': ['7 * enzyme * A', '- (1e-6 * P)']}
```

and this will generate the following Python function suitable for ODE solver

```
def P(t, y):
    exp_1 = (7 * y[0] * y[1])
    exp_2 = -(1e-6 * y[2])
    return exp_1 + exp_2
```

where `y[0]`, `y[1]`, and `y[2]` represent the concentration of enzyme, A, and P, respectively.

Hence, the availability of this script generator will allow for the generation of an executable Python ODE script file from the textual specification of a system of ODEs. This will enable ODE scripts to be generated during runtime.

As an example, given the following parameters:

- Rate of new births ($rate_{birth}$) = 0 % / day
- Rate of "zombie virus" transmission ($rate_{transmission}$) = 0.95 % / day
- Rate of natural human deaths ($rate_{death}$) = 0.01 % / day
- Rate of zombies resurrected from dead ($rate_{resurrect}$) = 0.02 % / day
- Rate of zombies destroyed by humans ($rate_{destroy}$) = 0.03 % / day
- Influx of 5 new humans per day ($rate_{influx}$)
- Maximum number of zombies = 700
- Minimum number of humans = 0

A system of ODEs describing zombie apocalypse (Munz et al., 2009) can be written as,

$$\begin{aligned}\frac{d\ human}{dt} &= influx + birth - zombied - death \\ &= \left(rate_{influx} + rate_{birth}\right) - \left(rate_{transmit} \times human \times zombie\right) \\ &\quad - \left(rate_{death} \times human\right) \\ &= (5.0 + 0.0) - (0.0095 \times human \times zombie) - (0.0001 \times human)\end{aligned}$$

$$\begin{aligned}\frac{d\ zombie}{dt} &= zombied + resurrected - destroyed \\ &= \left(rate_{transmit} \times human \times zombie\right) + \left(rate_{resurrect} \times dead\right) \\ &\quad - \left(rate_{destroy} \times human \times zombie\right) \\ &= (0.0095 \times human \times zombie) + (0.0002 \times dead) \\ &\quad - (0.0003 \times human \times zombie)\end{aligned}$$

$$\frac{d\,dead}{dt} = death + destroyed - resurrected$$
$$= (rate_{death} \times human) + (rate_{destroy} \times human \times zombie) - (rate_{resurrect} \times dead)$$
$$= (0.0001 \times human) + (0.0003 \times human \times zombie) - (0.0002 \times dead)$$

can be specified in the following format (file name = zombie_construct.py),

```
from ode import ODE_constructor

scriptfile = 'zombie_attack.py'
resultsfile = 'zombie_data.csv'
time = (0.0, 0.1, 100.0)
ODE_solver = 'RK4'
expressions = {'human': ['birth_rate',
                    '- (transmission_rate * human * zombie)',
                    '- (death_rate * human)'],
               'zombie': ['(transmission_rate * human *
                          zombie)',
                     '(resurrection_rate * dead)',
                     '- (destroy_rate * human * zombie)'],
               'dead': ['(death_rate * human)',
                        '(destroy_rate * human * zombie)',
                        '- (resurrection_rate * dead)']}
parameters = {'birth_rate': 0.0,
              'transmission_rate': 0.0095,
              'death_rate': 0.0001,
              'resurrection_rate': 0.0002,
              'destroy_rate':0.0003}
initial_conditions = {'human': 500.0,
                      'zombie': 0.0,
                      'dead': 0.0}
modifying_expressions = ['human = human + (5*step)']
lower_bound = {'human': [0.0, 0.0]}
upper_bound = {'zombie': [700, 700]}
overflow = 1e100
zerodivision = 1e100

s = ODE_constructor(scriptfile, resultsfile, time,
        ODE_solver, expressions, parameters,
        initial_conditions, modifying_expressions,
        lower_bound, upper_bound, overflow, zerodivision)
```

for simulating an initial population of 500 humans with no zombies (hence, initial "zombification" is spontaneous) over 100 days. The above specification will generate the following Python script file (file name = zombie_attack.py) when executed,

```
import ode

y = range(3)
y[0] = 0.0
y[1] = 0.0
y[2] = 500.0


def zombie(t, y):
    exp_1 = (0.0095 * y[2] * y[0])
    exp_2 = (0.0002 * y[1])
    exp_3 = - (0.0003 * y[2] * y[0])
    return exp_1 + exp_2 + exp_3
def dead(t, y):
    exp_1 = (0.0001 * y[2])
    exp_2 = (0.0003 * y[2] * y[0])
    exp_3 = - (0.0002 * y[1])
    return exp_1 + exp_2 + exp_3
def human(t, y):
    exp_1 = 0.0
    exp_2 = - (0.0095 * y[2] * y[0])
    exp_3 = - (0.0001 * y[2])
    return exp_1 + exp_2 + exp_3

def modifying_expression(y, step):
    y[2] = y[2] + (5*step)
    return y

ODE = range(3)
ODE[0] = zombie
ODE[1] = dead
ODE[2] = human

stime = 0.0
step = 0.1
etime = 100.0
lower_bound = {'2': [0.0, 0.0]}
upper_bound = {'0': [700, 700]}
overflow = 1e+100
zerodivision = 1e+100

f = open('zombie_data.csv', 'w')
f.write('time,zombie,dead,human' + '\n')

for x in ode.RK4(ODE, stime, y, step, etime,
        modifying_expression,
```

```
        lower_bound, upper_bound,
        overflow, zerodivision):
    f.write(','.join([str(item) for item in x]) + '\n')
f.close()
```

The above generated code is logically identical to the manually written ODE code script and produces identical simulation results,

```
import ode

birth = 0                   # birth rate
death = 0.0001              # natural death percent (per day)
transmission = 0.0095       # transmission percent  (per day)
resurect = 0.0002           # resurect percent (per day)
destroy = 0.0003            # destroy percent (per day)

def human(t, y):
    infected = transmission*y[0]*y[1]
    dead = death*y[0]
    return birth - infected - dead
def zombie(t, y):
    newly_infected = transmission*y[0]*y[1]
    resurrected = resurect*y[2]
    destroyed = destroy*y[0]*y[1]
    return newly_infected + resurrected - destroyed
def dead(t, y):
    natural_death = death*y[0]
    destroyed_zombies = destroy*y[0]*y[1]
    created_zombies = resurect*y[2]
    return natural_death + destroyed_zombies - \
           created_zombies

def modifying_expression(y, step):
    y[0] = y[0] + (5 * step)
    return y

lower_bound = {'2': [0.0, 0.0]}
upper_bound = {'0': [700, 700]}

f = [human, zombie, dead]   # system of ODEs

y = [500.0, 0, 0]  # initial human, zombie, death population
print('Solving using 4th order Runge Kutta method ......')
for i in [x for x in ode.RK4(f, 0.0, y, 0.1, 100.0,
          modifying_expression, lower_bound,
          upper_bound)]:
    print(','.join([str(z) for z in i]))
print('')
```

CODE FILE FOR ODE SOLVER (ODE.PY)

```
'''
Ordinary Differential Equation (ODE) Solvers.
Copyright (c) Maurice H.T. Ling <mauriceling@acm.org>
Date created: 20th December 2014'''
import re

def boundary_checker(y, boundary, type):
    '''Private function - called by ODE solvers to perform
    boundary checking of variable values and reset them to
    specific values if the original values fall out of the
    boundary values.

    Boundary parameter takes the form of a dictionary with
    variable number as key and a list of [<boundary value>,
    <value to set if boundary is exceeded>]. For example,
    The following dictionary for lower boundary (type =
    'lower') {'1': [0.0, 0.0], '5': [2.0, 2.0]} will set
    the lower boundary of variable y[0] and [5] to 0.0 and
    2.0 respectively. This also allows for setting to a
    different value - for example, {'1': [0.0, 1.0]} will
    set variable y[0] to 2.0 if the original y[0] value is
    negative.

    @param y: values for variables
    @type y: list
    @param boundary: set of values for boundary of variables
    @type boundary: dictionary
    @param type: the type of boundary to be checked, either
    'upper' (upper boundary) or 'lower' (lower boundary)
    '''
    for k in list(boundary.keys()):
        if y[int(k)] < boundary[k][0] and type == 'lower':
            y[int(k)] = boundary[k][1]
        if y[int(k)] > boundary[k][0] and type == 'upper':
            y[int(k)] = boundary[k][1]
    return y

def Euler(funcs, x0, y0, step, xmax, nonODEfunc=None,
          lower_bound=None, upper_bound=None,
          overflow=1e100, zerodivision=1e100):
    '''Generator to integrate a system of ODEs, y0' =
    funcs(x0, y0), using Euler method.

    A function (as nonODEfunc parameter) can be included to
    modify one or more variables (y0 list). This function
    will not be an ODE (not a dy/dt). This can be used to
    consolidate the modification of one or more variables
```

```
at each ODE solving step. For example, y[0] = y[1] /
y[2] can be written as

C{
def modifying_function(y, step):
    y[0] = y[1] / y[2]
    return y
}

This function must take 'y' (variable list) and 'step'
(time step) as parameters and must return 'y' (the
modified variable list). This function will execute
before boundary checking at each time step.

Upper and lower boundaries of one or more variable can
be set using upper_bound and lower_bound parameters
respectively. These parameters take the form of a
dictionary with variable number as key and a list
of [<boundary value>, <value to set if boundary is
exceeded>]. For example, the following dictionary for
lower boundary {'1': [0.0, 0.0], '5': [2.0, 2.0]} will
set the lower boundary of variable y[0] and y[5] to 0.0
and 2.0 respectively. This also allows for setting to
a different value - for example, {'1': [0.0, 1.0]} will
set variable y[0] to 2.0 if the original y[0] value is
negative.

@param funcs: system of differential equations
@type funcs: list
@param x0: initial value of x-axis, which is usually
starting time
@type x0: float
@param y0: initial values for variables
@type y0: list
@param step: step size on the x-axis (also known as step
in calculus)
@type step: float
@param xmax: maximum value of x-axis, which is usually
ending time
@type xmax: float
@param nonODEfunc: a function to modify the variable
list(y0)
@type nonODEfunc: function
@param lower_bound: set of values for lower boundary of
variables
@type lower_bound: dictionary
@param upper_bound: set of values for upper boundary of
variables
@type upper_bound: dictionary
@param overflow: value (usually a large value) to assign
in event of over flow error (usually caused by a large
```

```
    number) during integration. Default = 1e100.
    @type overflow: float
    @param zerodivision: value (usually a large value) to
    assign in event of zero division error, which results
    in positive infinity, during integration. Default =
    1e100.
    @type zerodivision: float
    '''
    yield [x0] + y0
    def solver(funcs, x0, y0, step):
        n = len(funcs)
        y1 = [0]*n
        for i in range(n):
            try: y1[i] = y0[i] + (step*funcs[i](x0, y0))
            except TypeError: pass
            except ZeroDivisionError: y0[i] = zerodivision
            except OverflowError: y0[i] = overflow
        return y1
    while x0 < xmax:
        y1 = solver(funcs, x0, y0, step)
        if nonODEfunc:
            y1 = nonODEfunc(y1, step)
        if lower_bound:
            y1 = boundary_checker(y1, lower_bound, 'lower')
        if upper_bound:
            y1 = boundary_checker(y1, upper_bound, 'upper')
        y0 = y1
        x0 = x0 + step
        yield [x0] + y0

def Heun(funcs, x0, y0, step, xmax, nonODEfunc=None,
         lower_bound=None, upper_bound=None,
         overflow=1e100, zerodivision=1e100):
    '''
    Generator to integrate a system of ODEs, y0' = funcs(x0,
    y0), using Heun's method, which is also known as Runge-
    Kutta 2nd method or Trapezoidal method.

    See Euler method documentation for description on
    parameters list.'''
    yield [x0] + y0
    def solver(funcs, x0, y0, step):
        n = len(funcs)
        f1 = [0]*n
        y1, y2 = [0]*n, [0]*n
        for i in range(n):
            try: f1[i] = funcs[i](x0, y0)
            except TypeError: pass
            except ZeroDivisionError: f1[i] = zerodivision
            except OverflowError: f1[i] = overflow
        for i in range(n):
```

```
            try: y1[i] = y0[i] + step*f1[i]
            except TypeError: pass
            except ZeroDivisionError: y1[i] = zerodivision
            except OverflowError: y2[i] = overflow
        for i in range(n):
            try: y2[i] = y0[i] + 0.5*step*(f1[i] + \
                                funcs[i](x0+step, y1))
            except TypeError: pass
            except ZeroDivisionError: y2[i] = zerodivision
            except OverflowError: y2[i] = overflow
        return y2
    while x0 < xmax:
        y2 = solver(funcs, x0, y0, step)
        if nonODEfunc:
            y1 = nonODEfunc(y1, step)
        if lower_bound:
            y2 = boundary_checker(y2, lower_bound, 'lower')
        if upper_bound:
            y2 = boundary_checker(y2, upper_bound, 'upper')
        y0 = y2
        x0 = x0 + step
        yield [x0] + y0

def RK3(funcs, x0, y0, step, xmax, nonODEfunc=None,
        lower_bound=None, upper_bound=None,
        overflow=1e100, zerodivision=1e100):
    '''
    Generator to integrate a system of ODEs, y0' = funcs(x0,
    y0), using third order Runge-Kutta method.

    See Euler method documentation for description on
    parameters list.'''
    yield [x0] + y0
    def solver(funcs, x0, y0, step):
        n = len(funcs)
        f1, f2, f3 = [0]*n, [0]*n, [0]*n
        y1, y2 = [0]*n, [0]*n
        y1 = [0]*n
        for i in range(n):
            try: f1[i] = step * funcs[i](x0, y0)
            except TypeError: pass
            except ZeroDivisionError: f1[i] = zerodivision
            except OverflowError: f1[i] = overflow
        for i in range(n):
            y1[i] = y0[i] + 0.5*f1[i]
        for i in range(n):
            try: f2[i] = step * funcs[i](x0 + 0.5*step, y1)
            except TypeError: pass
            except ZeroDivisionError: f2[i] = zerodivision
            except OverflowError: f2[i] = overflow
        for i in range(n):
```

```
                y1[i] = y0[i] - f1[i] + 2*f2[i]
            for i in range(n):
                try: f3[i] = step * funcs[i](x0 + step, y1)
                except TypeError: pass
                except ZeroDivisionError: f3[i] = zerodivision
                except OverflowError: f3[i] = overflow
            for i in range(n):
                try: y1[i] = y0[i] + (f1[i] + 4*f2[i] + \
                        f3[i])/6.0
                except TypeError: pass
                except ZeroDivisionError: y1[i] = zerodivision
                except OverflowError: y1[i] = overflow
            return y1
        while x0 < xmax:
            y1 = solver(funcs, x0, y0, step)
            if nonODEfunc:
                y1 = nonODEfunc(y1, step)
            if lower_bound:
                y1 = boundary_checker(y1, lower_bound, 'lower')
            if upper_bound:
                y1 = boundary_checker(y1, upper_bound, 'upper')
            y0 = y1
            x0 = x0 + step
            yield [x0] + y0

    def RK4(funcs, x0, y0, step, xmax, nonODEfunc=None,
            lower_bound=None, upper_bound=None,
            overflow=1e100, zerodivision=1e100):
        '''
        Generator to integrate a system of ODEs, y' = f(x, y),
        using fourth order Runge-Kutta method.

        See Euler method documentation for description on
        parameters list.'''
        yield [x0] + y0
        def solver(funcs, x0, y0, step):
            n = len(funcs)
            f1, f2, f3, f4 = [0]*n, [0]*n, [0]*n, [0]*n
            y1 = [0]*n
            for i in range(n):
                try: f1[i] = funcs[i](x0, y0)
                except TypeError: pass
                except ZeroDivisionError: f1[i] = zerodivision
                except OverflowError: f1[i] = overflow
            for j in range(n):
                y1[j] = y0[j] + (0.5*step*f1[j])
            for i in range(n):
                try: f2[i] = funcs[i]((x0+(0.5*step)), y1)
                except TypeError: pass
                except ZeroDivisionError: f2[i] = zerodivision
                except OverflowError: f2[i] = overflow
```

```
        for j in range(n):
            y1[j] = y0[j] + (0.5*step*f2[j])
        for i in range(n):
            try: f3[i] = funcs[i]((x0+(0.5*step)), y1)
            except TypeError: pass
            except ZeroDivisionError: f3[i] = zerodivision
            except OverflowError: f3[i] = overflow
        for j in range(n):
            y1[j] = y0[j] + (step*f3[j])
        for i in range(n):
            try: f4[i] = funcs[i]((x0+step), y1)
            except TypeError: pass
            except ZeroDivisionError: f4[i] = zerodivision
            except OverflowError: f4[i] = overflow
        for i in range(n):
            try: y1[i] = y0[i] + (step * \
                    (f1[i] + (2.0*f2[i]) + (2.0*f3[i]) + \
                f4[i]) / 6.0)
            except TypeError: pass
            except ZeroDivisionError: y1[i] = zerodivision
            except OverflowError: y1[i] = overflow
        return y1
    while x0 < xmax:
        y1 = solver(funcs, x0, y0, step)
        if nonODEfunc:
            y1 = nonODEfunc(y1, step)
        if lower_bound:
            y1 = boundary_checker(y1, lower_bound, 'lower')
        if upper_bound:
            y1 = boundary_checker(y1, upper_bound, 'upper')
        y0 = y1
        x0 = x0 + step
        yield [x0] + y0

def RK38(funcs, x0, y0, step, xmax, nonODEfunc=None,
        lower_bound=None, upper_bound=None,
        overflow=1e100, zerodivision=1e100):
    '''
    Generator to integrate a system of ODEs, y' = f(x, y),
    using fourth order Runge-Kutta method, 3/8 rule.

    See Euler method documentation for description on
    parameters list.'''
    yield [x0] + y0
    def solver(funcs, x0, y0, step):
        n = len(funcs)
        f1, f2, f3, f4 = [0]*n, [0]*n, [0]*n, [0]*n
        y1 = [0]*n
        for i in range(n):
            try: f1[i] = funcs[i](x0, y0)
            except TypeError: pass
```

```
            except ZeroDivisionError: f1[i] = zerodivision
            except OverflowError: f1[i] = overflow
        for j in range(n):
            y1[j] = y0[j] + (step*f1[j]/3.0)
        for i in range(n):
            try: f2[i] = funcs[i]((x0+(step/3.0)), y1)
            except TypeError: pass
            except ZeroDivisionError: f2[i] = zerodivision
            except OverflowError: f2[i] = overflow
        for j in range(n):
            y1[j] = y0[j] + (-1*step*f1[j]/3.0) + \
                    (step*f2[j])
        for i in range(n):
            try: f3[i] = funcs[i]((x0+(2*step/3.0)), y1)
            except TypeError: pass
            except ZeroDivisionError: f3[i] = zerodivision
            except OverflowError: f3[i] = overflow
        for j in range(n):
            y1[j] = y0[j] + (step*f1[j]) + \
                   (-step*f2[j]) + (step*f3[j])
        for i in range(n):
            try: f4[i] = funcs[i]((x0+step), y1)
            except TypeError: pass
            except ZeroDivisionError: f4[i] = zerodivision
            except OverflowError: f4[i] = overflow
        for i in range(n):
            try: y1[i] = y0[i] + (step * \
                    (f1[i] + (3.0*f2[i]) + (3.0*f3[i]) + \
                    f4[i]) / 8.0)
            except TypeError: pass
            except ZeroDivisionError: y1[i] = zerodivision
            except OverflowError: y1[i] = overflow
        return y1
    while x0 < xmax:
        y1 = solver(funcs, x0, y0, step)
        if nonODEfunc:
            y1 = nonODEfunc(y1, step)
        if lower_bound:
            y1 = boundary_checker(y1, lower_bound, 'lower')
        if upper_bound:
            y1 = boundary_checker(y1, upper_bound, 'upper')
        y0 = y1
        x0 = x0 + step
        yield [x0] + y0

def CK4(funcs, x0, y0, step, xmax, nonODEfunc=None,
        lower_bound=None, upper_bound=None,
        overflow=1e100, zerodivision=1e100):
    '''
    Generator to integrate a system of ODEs, y' = f(x, y),
    using fifth order Cash-Karp method.
```

```
See Euler method documentation for description on
parameters list.'''
yield [x0] + y0
def solver(funcs, x0, y0, step):
    n = len(funcs)
    f1, f2, f3, f4, f5, f6 = [0]*n, [0]*n, [0]*n, \
                             [0]*n, [0]*n, [0]*n
    y1 = [0]*n
    for i in range(n):
        try: f1[i] = funcs[i](x0, y0)
        except TypeError: pass
        except ZeroDivisionError: f1[i] = zerodivision
        except OverflowError: f1[i] = overflow
    for j in range(n):
        y1[j] = y0[j] + (0.2*step*f1[j])
    for i in range(n):
        try: f2[i] = funcs[i]((x0+(0.2*step)), y1)
        except TypeError: pass
        except ZeroDivisionError: f2[i] = zerodivision
        except OverflowError: f2[i] = overflow
    for j in range(n):
        y1[j] = y0[j] + (0.075*step*f1[j]) + \
                (0.225*step*f2[j])
    for i in range(n):
        try: f3[i] = funcs[i]((x0+(0.3*step)), y1)
        except TypeError: pass
        except ZeroDivisionError: f3[i] = zerodivision
        except OverflowError: f3[i] = overflow
    for j in range(n):
        y1[j] = y0[j] + (0.3*step*f1[j]) + \
                (-0.9*step*f2[j]) + (1.2*step*f3[j])
    for i in range(n):
        try: f4[i] = funcs[i]((x0+(0.6*step)), y1)
        except TypeError: pass
        except ZeroDivisionError: f4[i] = zerodivision
        except OverflowError: f4[i] = overflow
    for j in range(n):
        y1[j] = y0[j] + (-11*step*f1[j]/54.0) + \
            (2.5*step*f2[j]) + (-70*step*f3[j]/27.0) + \
            (35*step*f4[j]/27.0)
    for i in range(n):
        try: f5[i] = funcs[i](x0+step, y1)
        except TypeError: pass
        except ZeroDivisionError: f5[i] = zerodivision
        except OverflowError: f5[i] = overflow
    for j in range(n):
        y1[j] = y0[j] + (1631*step*f1[j]/55296) + \
            (175*step*f2[j]/512) + \
            (575*step*f3[j]/13824) \
            + (44275*step*f4[j]/110592) + \
            (253*step*f5[j]/4096)
```

```
        for i in range(n):
            try: f6[i] = funcs[i](x0+(0.875*step), y1)
            except TypeError: pass
            except ZeroDivisionError: f6[i] = zerodivision
            except OverflowError: f6[i] = overflow
        for i in range(n):
            try: y1[i] = y0[i] + (step * \
                    ((2825*f1[i]/27648) + \
                     (18575*f3[i]/48384) + \
                     (13525*f4[i]/55296) + \
                     (277*f5[i]/14336) + \
                     (0.25*f6[i])))
            except TypeError: pass
            except ZeroDivisionError: y1[i] = zerodivision
            except OverflowError: y1[i] = overflow
        return y1
    while x0 < xmax:
        y1 = solver(funcs, x0, y0, step)
        if nonODEfunc:
            y1 = nonODEfunc(y1, step)
        if lower_bound:
            y1 = boundary_checker(y1, lower_bound, 'lower')
        if upper_bound:
            y1 = boundary_checker(y1, upper_bound, 'upper')
        y0 = y1
        x0 = x0 + step
        yield [x0] + y0

def CK5(funcs, x0, y0, step, xmax, nonODEfunc=None,
        lower_bound=None, upper_bound=None,
        overflow=1e100, zerodivision=1e100):
    '''
    Generator to integrate a system of ODEs, y' = f(x, y),
    using fifth order Cash-Karp method.

    See Euler method documentation for description on
    parameters list.'''
    yield [x0] + y0
    def solver(funcs, x0, y0, step):
        n = len(funcs)
        f1, f2, f3, f4, f5, f6 = [0]*n, [0]*n, [0]*n, \
                                 [0]*n, [0]*n, [0]*n
        y1 = [0]*n
        for i in range(n):
            try: f1[i] = funcs[i](x0, y0)
            except TypeError: pass
            except ZeroDivisionError: f1[i] = zerodivision
            except OverflowError: f1[i] = overflow
        for j in range(n):
            y1[j] = y0[j] + (0.2*step*f1[j])
        for i in range(n):
```

```
        try: f2[i] = funcs[i]((x0+(0.2*step)), y1)
        except TypeError: pass
        except ZeroDivisionError: f2[i] = zerodivision
        except OverflowError: f2[i] = overflow
    for j in range(n):
        y1[j] = y0[j] + (0.075*step*f1[j]) + \
                (0.225*step*f2[j])
    for i in range(n):
        try: f3[i] = funcs[i]((x0+(0.3*step)), y1)
        except TypeError: pass
        except ZeroDivisionError: f3[i] = zerodivision
        except OverflowError: f3[i] = overflow
    for j in range(n):
        y1[j] = y0[j] + (0.3*step*f1[j]) + \
            (-0.9*step*f2[j]) + (1.2*step*f3[j])
    for i in range(n):
        try: f4[i] = funcs[i]((x0+(0.6*step)), y1)
        except TypeError: pass
        except ZeroDivisionError: f4[i] = zerodivision
        except OverflowError: f4[i] = overflow
    for j in range(n):
        y1[j] = y0[j] + (-11*step*f1[j]/54.0) + \
          (2.5*step*f2[j]) + (-70*step*f3[j]/27.0) + \
          (35*step*f4[j]/27.0)
    for i in range(n):
        try: f5[i] = funcs[i](x0+step, y1)
        except TypeError: pass
        except ZeroDivisionError: f5[i] = zerodivision
        except OverflowError: f5[i] = overflow
    for j in range(n):
        y1[j] = y0[j] + (1631*step*f1[j]/55296.0) + \
                (175*step*f2[j]/512.0) + \
                (575*step*f3[j]/13824.0) + \
                (44275*step*f4[j]/110592.0) + \
                (253*step*f5[j]/4096.0)
    for i in range(n):
        try: f6[i] = funcs[i](x0+(0.875*step), y1)
        except TypeError: pass
        except ZeroDivisionError: f6[i] = zerodivision
        except OverflowError: f6[i] = overflow
    for i in range(n):
        try: y1[i] = y0[i] + (step * \
                ((37*f1[i]/378.0) + \
                 (250*f3[i]/621.0) + \
                 (125*f4[i]/594.0) + \
                 (512*f6[i]/1771.0)))
        except TypeError: pass
        except ZeroDivisionError: y1[i] = zerodivision
        except OverflowError: y1[i] = overflow
    return y1
while x0 < xmax:
```

```
        y1 = solver(funcs, x0, y0, step)
        if nonODEfunc:
            y1 = nonODEfunc(y1, step)
        if lower_bound:
            y1 = boundary_checker(y1, lower_bound, 'lower')
        if upper_bound:
            y1 = boundary_checker(y1, upper_bound, 'upper')
        y0 = y1
        x0 = x0 + step
        yield [x0] + y0

def RKF4(funcs, x0, y0, step, xmax, nonODEfunc=None,
        lower_bound=None, upper_bound=None,
        overflow=1e100, zerodivision=1e100):
    '''
    Generator to integrate a system of ODEs, y' = f(x, y),
    using fourth order Runge-Kutta_Fehlberg method.

    See Euler method documentation for description on
    parameters list.'''
    yield [x0] + y0
    def solver(funcs, x0, y0, step):
        n = len(funcs)
        f1, f2, f3, f4, f5, f6 = [0]*n, [0]*n, [0]*n, \
                                 [0]*n, [0]*n, [0]*n
        y1 = [0]*n
        for i in range(n):
            try: f1[i] = funcs[i](x0, y0)
            except TypeError: pass
            except ZeroDivisionError: f1[i] = zerodivision
            except OverflowError: f1[i] = overflow
        for j in range(n):
            y1[j] = y0[j] + (0.25*step*f1[j])
        for i in range(n):
            try: f2[i] = funcs[i]((x0+(0.25*step)), y1)
            except TypeError: pass
            except ZeroDivisionError: f2[i] = zerodivision
            except OverflowError: f2[i] = overflow
        for j in range(n):
            y1[j] = y0[j] + (3*step*f1[j]/32.0) + \
                    (9*step*f2[j]/32.0)
        for i in range(n):
            try: f3[i] = funcs[i]((x0+(3*step/8.0)), y1)
            except TypeError: pass
            except ZeroDivisionError: f3[i] = zerodivision
            except OverflowError: f3[i] = overflow
        for j in range(n):
            y1[j] = y0[j] + (1932*step*f1[j]/2197.0) + \
                    (-7200*step*f2[j]/2197.0) + \
                    (7296*step*f3[j]/2197.0)
        for i in range(n):
```

```
            try: f4[i] = funcs[i]((x0+(12*step/13.0)), y1)
            except TypeError: pass
            except ZeroDivisionError: f4[i] = zerodivision
            except OverflowError: f4[i] = overflow
        for j in range(n):
            y1[j] = y0[j] + (439*step*f1[j]/216.0) + \
                    (-8.0*step*f2[j]) + \
                    (3680*step*f3[j]/513.0) + \
                    (-845*step*f4[j]/4104.0)
        for i in range(n):
            try: f5[i] = funcs[i](x0+step, y1)
            except TypeError: pass
            except ZeroDivisionError: f5[i] = zerodivision
            except OverflowError: f5[i] = overflow
        for j in range(n):
            y1[j] = y0[j] + (-8*step*f1[j]/27.0) + \
                    (2.0*step*f2[j]) + \
                    (-3544*step*f3[j]/2565.0) + \
                    (1859*step*f4[j]/4104.0) + \
                    (-11*step*f5[j]/40.0)
        for i in range(n):
            try: f6[i] = funcs[i](x0+(0.5*step), y1)
            except TypeError: pass
            except ZeroDivisionError: f6[i] = zerodivision
            except OverflowError: f6[i] = overflow
        for i in range(n):
            try: y1[i] = y0[i] + (step * \
                         ((25*f1[i]/216.0) + \
                         (1408*f3[i]/2565.0) + \
                         (2197*f4[i]/4104.0) + \
                         (-0.2*f5[i])))
            except TypeError: pass
            except ZeroDivisionError: y1[i] = zerodivision
            except OverflowError: y1[i] = overflow
        return y1
    while x0 < xmax:
        y1 = solver(funcs, x0, y0, step)
        if nonODEfunc:
            y1 = nonODEfunc(y1, step)
        if lower_bound:
            y1 = boundary_checker(y1, lower_bound, 'lower')
        if upper_bound:
            y1 = boundary_checker(y1, upper_bound, 'upper')
        y0 = y1
        x0 = x0 + step
        yield [x0] + y0

def RKF5(funcs, x0, y0, step, xmax, nonODEfunc=None,
        lower_bound=None, upper_bound=None,
        overflow=1e100, zerodivision=1e100):
    '''
```

```
    Generator to integrate a system of ODEs, y' = f(x, y),
    using fifth order Runge-Kutta_Fehlberg method.

    See Euler method documentation for description on
    parameters list.'''
    yield [x0] + y0
    def solver(funcs, x0, y0, step):
        n = len(funcs)
        f1, f2, f3, f4, f5, f6 = [0]*n, [0]*n, [0]*n, \
                                 [0]*n, [0]*n, [0]*n
        y1 = [0]*n
        for i in range(n):
            try: f1[i] = funcs[i](x0, y0)
            except TypeError: pass
            except ZeroDivisionError: f1[i] = zerodivision
            except OverflowError: f1[i] = overflow
        for j in range(n):
            y1[j] = y0[j] + (0.25*step*f1[j])
        for i in range(n):
            try: f2[i] = funcs[i]((x0+(0.25*step)), y1)
            except TypeError: pass
            except ZeroDivisionError: f2[i] = zerodivision
            except OverflowError: f2[i] = overflow
        for j in range(n):
            y1[j] = y0[j] + (3*step*f1[j]/32.0) + \
                    (9*step*f2[j]/32.0)
        for i in range(n):
            try: f3[i] = funcs[i]((x0+(3*step/8.0)), y1)
            except TypeError: pass
            except ZeroDivisionError: f3[i] = zerodivision
            except OverflowError: f3[i] = overflow
        for j in range(n):
            y1[j] = y0[j] + (1932*step*f1[j]/2197.0) + \
                     (-7200*step*f2[j]/2197.0) + \
                     (7296*step*f3[j]/2197.0)
        for i in range(n):
            try: f4[i] = funcs[i]((x0+(12*step/13.0)), y1)
            except TypeError: pass
            except ZeroDivisionError: f4[i] = zerodivision
            except OverflowError: f4[i] = overflow
        for j in range(n):
            y1[j] = y0[j] + (439*step*f1[j]/216.0) + \
                     (-8.0*step*f2[j]) + \
                     (3680*step*f3[j]/513.0) + \
                     (-845*step*f4[j]/4104.0)
        for i in range(n):
            try: f5[i] = funcs[i](x0+step, y1)
            except TypeError: pass
            except ZeroDivisionError: f5[i] = zerodivision
            except OverflowError: f5[i] = overflow
        for j in range(n):
```

```
        y1[j] = y0[j] + (-8*step*f1[j]/27.0) + \
                (2.0*step*f2[j]) + \
                (-3544*step*f3[j]/2565.0) + \
                (1859*step*f4[j]/4104.0) + \
                (-11*step*f5[j]/40.0)
    for i in range(n):
        try: f6[i] = funcs[i](x0+(0.5*step), y1)
        except TypeError: pass
        except ZeroDivisionError: f6[i] = zerodivision
        except OverflowError: f6[i] = overflow
    for i in range(n):
        try: y1[i] = y0[i] + (step * \
                     ((16*f1[i]/135.0) + \
                      (6656*f3[i]/12825.0) + \
                      (28561*f4[i]/56430.0) + \
                      (-9*f5[i]/50) + \
                      (2*f6[i]/55)))
        except TypeError: pass
        except ZeroDivisionError: y1[i] = zerodivision
        except OverflowError: y1[i] = overflow
    return y1
while x0 < xmax:
    y1 = solver(funcs, x0, y0, step)
    if nonODEfunc:
        y1 = nonODEfunc(y1, step)
    if lower_bound:
        y1 = boundary_checker(y1, lower_bound, 'lower')
    if upper_bound:
        y1 = boundary_checker(y1, upper_bound, 'upper')
    y0 = y1
    x0 = x0 + step
    yield [x0] + y0

def DP4(funcs, x0, y0, step, xmax, nonODEfunc=None,
    lower_bound=None, upper_bound=None,
    overflow=1e100, zerodivision=1e100):
    '''
    Generator to integrate a system of ODEs, y' = f(x, y),
    using fourth order Dormand-Prince method.

    See Euler method documentation for description on
    parameters list.'''
    yield [x0] + y0
    def solver(funcs, x0, y0, step):
        n = len(funcs)
        f1, f2, f3 = [0]*n, [0]*n, [0]*n
        f4, f5, f6, f7 = [0]*n, [0]*n, [0]*n, [0]*n
        y1 = [0]*n
        for i in range(n):
            try: f1[i] = funcs[i](x0, y0)
            except TypeError: pass
```

```
        except ZeroDivisionError: f1[i] = zerodivision
        except OverflowError: f1[i] = overflow
    for j in range(n):
        y1[j] = y0[j] + (0.2*step*f1[j])
    for i in range(n):
        try: f2[i] = funcs[i]((x0+(0.2*step)), y1)
        except TypeError: pass
        except ZeroDivisionError: f2[i] = zerodivision
        except OverflowError: f2[i] = overflow
    for j in range(n):
        y1[j] = y0[j] + (3*step*f1[j]/40.0) + \
                (9*step*f2[j]/40.0)
    for i in range(n):
        try: f3[i] = funcs[i]((x0+(0.3*step)), y1)
        except TypeError: pass
        except ZeroDivisionError: f3[i] = zerodivision
        except OverflowError: f3[i] = overflow
    for j in range(n):
        y1[j] = y0[j] + (44*step*f1[j]/45.0) + \
           (-56*step*f2[j]/15.0) + (32*step*f3[j]/9.0)
    for i in range(n):
        try: f4[i] = funcs[i]((x0+(0.8*step)), y1)
        except TypeError: pass
        except ZeroDivisionError: f4[i] = zerodivision
        except OverflowError: f4[i] = overflow
    for j in range(n):
        y1[j] = y0[j] + (19372*step*f1[j]/6561.0) + \
                (-25360*step*f2[j]/2187.0) + \
                (64448*step*f3[j]/6561.0) + \
                (-212*step*f4[j]/729.0)
    for i in range(n):
        try: f5[i] = funcs[i](x0+(8*step/9.0), y1)
        except TypeError: pass
        except ZeroDivisionError: f5[i] = zerodivision
        except OverflowError: f5[i] = overflow
    for j in range(n):
        y1[j] = y0[j] + (9017*step*f1[j]/3168.0) + \
                (-355*step*f2[j]/33.0) + \
                (46732*step*f3[j]/5247.0) + \
                (49*step*f4[j]/176.0) + \
                (-5103*step*f5[j]/18656.0)
    for i in range(n):
        try: f6[i] = funcs[i](x0+step, y1)
        except TypeError: pass
        except ZeroDivisionError: f6[i] = zerodivision
        except OverflowError: f6[i] = overflow
    for j in range(n):
        y1[j] = y0[j] + (35*step*f1[j]/384.0) + \
                (500*step*f3[j]/1113.0) + \
                (125*step*f4[j]/192.0) + \
                (-2187*step*f5[j]/6784.0) + \
```

```
                    (11*step*f6[j]/84.0)
        for i in range(n):
            try: f7[i] = funcs[i](x0+step, y1)
            except TypeError: pass
            except ZeroDivisionError: f7[i] = zerodivision
            except OverflowError: f7[i] = overflow
        for i in rangc(n):
            try: y1[i] = y0[i] + (step * \
                         ((5179*f1[i]/57600.0) + \
                          (7571*f3[i]/16695.0) + \
                          (393*f4[i]/640.0) + \
                          (-92097*f5[i]/339200.0) + \
                          (187*f6[i]/2100.0) + \
                          (f7[i]/40.0)))
            except TypeError: pass
            except ZeroDivisionError: y1[i] = zerodivision
            except OverflowError: y1[i] = overflow
        return y1
    while x0 < xmax:
        y1 = solver(funcs, x0, y0, step)
        if nonODEfunc:
            y1 = nonODEfunc(y1, step)
        if lower_bound:
            y1 = boundary_checker(y1, lower_bound, 'lower')
        if upper_bound:
            y1 = boundary_checker(y1, upper_bound, 'upper')
        y0 = y1
        x0 = x0 + step
        yield [x0] + y0

def DP5(funcs, x0, y0, step, xmax, nonODEfunc=None,
        lower_bound=None, upper_bound=None,
        overflow=1e100, zerodivision=1e100):
    '''
    Generator to integrate a system of ODEs, y' = f(x, y),
    using fifth order Dormand-Prince method.

    See Euler method documentation for description on
    parameters list.'''
    yield [x0] + y0
    def solver(funcs, x0, y0, step):
        n = len(funcs)
        f1, f2, f3 = [0]*n, [0]*n, [0]*n
        f4, f5, f6, f7 = [0]*n, [0]*n, [0]*n, [0]*n
        y1 = [0]*n
        for i in range(n):
            try: f1[i] = funcs[i](x0, y0)
            except TypeError: pass
            except ZeroDivisionError: f1[i] = zerodivision
            except OverflowError: f1[i] = overflow
        for j in range(n):
```

```
        y1[j] = y0[j] + (0.2*step*f1[j])
    for i in range(n):
        try: f2[i] = funcs[i]((x0+(0.2*step)), y1)
        except TypeError: pass
        except ZeroDivisionError: f2[i] = zerodivision
        except OverflowError: f2[i] = overflow
    for j in range(n):
        y1[j] = y0[j] + (3*step*f1[j]/40.0) + \
                (9*step*f2[j]/40.0)
    for i in range(n):
        try: f3[i] = funcs[i]((x0+(0.3*step)), y1)
        except TypeError: pass
        except ZeroDivisionError: f3[i] = zerodivision
        except OverflowError: f3[i] = overflow
    for j in range(n):
        y1[j] = y0[j] + (44*step*f1[j]/45.0) + \
          (-56*step*f2[j]/15.0) + (32*step*f3[j]/9.0)
    for i in range(n):
        try: f4[i] = funcs[i]((x0+(0.8*step)), y1)
        except TypeError: pass
        except ZeroDivisionError: f4[i] = zerodivision
        except OverflowError: f4[i] = overflow
    for j in range(n):
        y1[j] = y0[j] + (19372*step*f1[j]/6561.0) + \
                (-25360*step*f2[j]/2187.0) + \
                (64448*step*f3[j]/6561.0) + \
                (-212*step*f4[j]/729.0)
    for i in range(n):
        try: f5[i] = funcs[i](x0+(8*step/9.0), y1)
        except TypeError: pass
        except ZeroDivisionError: f5[i] = zerodivision
        except OverflowError: f5[i] = overflow
    for j in range(n):
        y1[j] = y0[j] + (9017*step*f1[j]/3168.0) + \
                (-355*step*f2[j]/33.0) + \
                (46732*step*f3[j]/5247.0) + \
                (49*step*f4[j]/176.0) + \
                (-5103*step*f5[j]/18656.0)
    for i in range(n):
        try: f6[i] = funcs[i](x0+step, y1)
        except TypeError: pass
        except ZeroDivisionError: f6[i] = zerodivision
        except OverflowError: f6[i] = overflow
    for j in range(n):
        y1[j] = y0[j] + (35*step*f1[j]/384.0) + \
                (500*step*f3[j]/1113.0) + \
                (125*step*f4[j]/192.0) + \
                (-2187*step*f5[j]/6784.0) + \
                (11*step*f6[j]/84.0)
    for i in range(n):
        try: f7[i] = funcs[i](x0+step, y1)
```

```
            except TypeError: pass
            except ZeroDivisionError: f7[i] = zerodivision
            except OverflowError: f7[i] = overflow
        for i in range(n):
            try: y1[i] = y0[i] + (step * \
                         ((35*f1[i]/384.0) + \
                          (500*f3[i]/1113.0) + \
                          (125*f4[i]/192.0) + \
                          (-2187*f5[i]/6784.0) + \
                          (11*f6[i]/84.0)))
            except TypeError: pass
            except ZeroDivisionError: y1[i] = zerodivision
            except OverflowError: y1[i] = overflow
        return y1
    while x0 < xmax:
        y1 = solver(funcs, x0, y0, step)
        if nonODEfunc:
            y1 = nonODEfunc(y1, step)
        if lower_bound:
            y1 = boundary_checker(y1, lower_bound, 'lower')
        if upper_bound:
            y1 = boundary_checker(y1, upper_bound, 'upper')
        y0 = y1
        x0 = x0 + step
        yield [x0] + y0

def _equation_constructor(expressions={},
                          parameters={},
                          variables=[]):
    '''
    Private function to support ODE_constructor to generate
    ODE function for each ODE.

    @param expressions: dictionary of expressions for
    ODE(s).
    Please see above documentation.
    @param parameters: dictionary of parameter values to be
    substituted into the ODE equations
    @param variables: list of additional variables to be
    tagged for substitution(s)
    @return: tuple of (<list of generated ODE functions>,
    <list of tagged variables>) '''
    statements = []
    # Generate ODE function, one at a time
    for name in expressions.keys():
        # Generate ODE function definition
        stmt = '\ndef %s(t, y):' % str(name)
        expression = expressions[name]
        if type(expression) == type(''):
            expression = [expression]
        count = 1
```

```
        # Generate list of expression(s) for current ODE
        exp_list = []
        variables = list(set(variables + \
                             expressions.keys()))
        for exp in expression:
            # Substitute parameter/variable values
            for k in parameters.keys():
                exp = exp.replace(str(k),
                                  str(parameters[k]))
            for v in variables:
                exp = exp.replace(str(v), 'v@'+str(v))
            # Generate expression codes
            stmt = stmt + '\n    exp_%s = %s' % (str(count),
                                                 str(exp))
            exp_list.append('exp_%s' % str(count))
            count = count + 1
        # Compile generated expressions into return value
        return_stmt = '\n    return ' + ' + '.join(exp_list)
        stmt = stmt + return_stmt
        statements.append(stmt)
    return (statements, variables)

def _modifying_constructor(modifying_expressions):
    '''Private function to support ODE_constructor to
    Generate function code for modifying expressions.

    @param modifying_expressions: list of expressions to
    modify the variables
    @return: generated function code'''
    stmt = '\ndef modifying_expression(y, step):'
    for exp in modifying_expressions:
        stmt = stmt + '\n    %s' % str(exp)
    stmt = stmt + '\n    return y'
    return stmt

def ODE_constructor(scriptfile,
                    resultsfile,
                    time=(0.0, 0.1, 100.0),
                    ODE_solver='RK4',
                    expressions={},
                    parameters={},
                    initial_conditions={},
                    modifying_expressions=[],
                    lower_bound=None, upper_bound=None,
                    overflow=1e100, zerodivision=1e100):
    '''Function to construct an ODE simulation script file
    from given definitions.

    For example, the following system of ODEs

    M{
```

```
d(human)/dt = birth - zombied - death
d(zombie)/dt = zombied + resurrected - destroyed
d(dead)/dt = death + destroyed - resurrected

birth = 0
zombied = 0.0095 * human * zombie
death = 0.0001 * human
resurrected = 0.0002 * dead
destroyed = 0.0003 * human * zombie

where initially (t0),

number of humans = 500
number of zombies = 0
number of dead = 0

and with the following boundaries

number of humans > 0
number of zombies < 10000
number of dead < 10000

and there is an influx of 5 humans per day into the
infected village}

can be specified as

C{
scriptfile = 'zombie_attack.py'
resultsfile = 'zombie_data.csv'
time = (0.0, 0.1, 100.0)
ODE_solver = 'RK4'
expressions = \
{'human': ['birth_rate',
           '- (transmission_rate * human * zombie)',
           '- (death_rate * human)'],
 'zombie': ['(transmission_rate * human * zombie)',
            '(resurrection_rate * dead)',
            '- (destroy_rate * human * zombie)'],
 'dead': ['(death_rate * human)',
          '(destroy_rate * human * zombie)',
          '- (resurrection_rate * dead)']}
parameters = {'birth_rate': 0.0,
              'transmission_rate': 0.0095,
              'death_rate': 0.0001,
              'resurrection_rate': 0.0002,
              'destroy_rate':0.0003}
initial_conditions = {'human': 500.0,
                      'zombie': 0.0,
                      'dead': 0.0}
modifying_expression = ['human = human + (5 * step)']
```

```
lower_bound = {'human': [0.0, 0.0]}
upper_bound = {'zombie': [10000.0, 10000.0],
               'dead': [10000.0, 10000.0]}
overflow = 1e100
zerodivision = 1e100
}

@param scriptfile: name of Python file for the generated
ODE script file
@type scriptfile: string
@param resultsfile: name of ODE simulation results file
(CSV file), which will be included into the generated
simulation script file
@type resultsfile: string
@param time: tuple of time parameters for simulation in
the format of (<start time>, <time step>, <end time>).
Default = (0.0, 0.1, 100.0)
@param ODE_solver: name of ODE solver to use. Default
= RK4
@param expressions: dictionary of expressions for
ODE(s). Please see above documentation.
@param parameters: dictionary of parameter values to be
substituted into the ODE equations
@param modifying_expressions: list of expressions to
modify the variables
@param initial_conditions: dictionary of initial
conditions for each ODE
@param lower_bound: set of values for lower boundary of
variables
@type lower_bound: dictionary
@param upper_bound: set of values for upper boundary of
variables
@type upper_bound: dictionary
@param overflow: value (usually a large value) to assign
in event of over flow error (usually caused by a large
number) during integration. Default = 1e100.
@type overflow: float
@param zerodivision: value (usually a large value) to
assign in event of zero division error, which results
in positive infinity, during integration. Default =
1e100.
@type zerodivision: float
@return: generated ODE codes for the entire script
@rtype: list'''
statements = []
initial_conditions_list = initial_conditions.keys()
# Construct ODE functions
(ODE_functions, variables) = \
    _equation_constructor(expressions, parameters,
                          initial_conditions_list)
# Construct modifying expression
```

```
mexpression = \
    _modifying_constructor(modifying_expressions)
# Generate ODE functions/equations table
table = {}
count = 0
for v in initial_conditions_list:
    table[str(v)] = str(count)
    count = count + 1
# Generate lower boundary table
if lower_bound != None:
    lbound = {}
    for k in lower_bound.keys():
        lbound[table[k]] = lower_bound[k]
else:
    lbound = None
# Generate upper boundary table
if upper_bound != None:
    ubound = {}
    for k in upper_bound.keys():
        ubound[table[k]] = upper_bound[k]
else:
    ubound = None
# Write ode module import
statements.append('import ode\n\n')
# Generate variable array and statements
statements.append('y = range(%s)\n' % \
                  str(len(variables)))
count = 0
for k in initial_conditions_list:
    statements.append('y[%s] = %s\n' % \
             (str(count), str(initial_conditions[k])))
    count = count + 1
statements.append('\n')
# Perform variable replacements in ODE functions, and
# write functions
for funct in ODE_functions:
    for v in table.keys():
        funct = funct.replace('v@'+v, 'y[%s]' % \
                                  table[v])
    statements.append(funct)
statements.append('\n\n')
# Perform variable replacements in modifying
# expression, and write out
for v in table.keys():
    mexpression = mexpression.replace(v, 'y[%s]' % \
                                          table[v])
statements.append(mexpression)
statements.append('\n\n')
# Generate ODE assignment array and statements
statements.append('ODE = range(%s)\n' % \
                      str(len(variables)))
```

```
    count = 0
    for k in initial_conditions_list:
        statements.append('ODE[%s] = %s\n' % \
                          (str(count), str(k)))
        count = count + 1
    statements.append('\n')
    # Generate ODE execution codes
    statements.append('stime = %s \n' % str(time[0]))
    statements.append('step = %s \n' % str(time[1]))
    statements.append('etime = %s \n' % str(time[2]))
    statements.append('lower_bound = %s \n' % str(lbound))
    statements.append('upper_bound = %s \n' % str(ubound))
    statements.append('overflow = %s \n' % str(overflow))
    statements.append('zerodivision = %s \n' % \
                      str(zerodivision))
    statements.append('\n')
    statements.append("f = open('%s', 'w')\n" % \
                      str(resultsfile))
    headers = ['time'] + \
              [str(k) for k in initial_conditions.keys()]
    headers = ','.join(headers)
    statements.append("f.write('%s' + '\\n') \n" % headers)
    statements.append('\n')
    statements.append('for x in ode.%s(ODE, stime, y, \n'\
        % str(ODE_solver))
    statements.append('    step, etime, \n')
    statements.append('    modifying_expression, \n')
    statements.append('    lower_bound, upper_bound, \n')
    statements.append('    overflow, zerodivision): \n')
    statements.append("    f.write(','.join([str(item) \n")
    statements.append("           for item in x]) + '\\n')")
    statements.append('\n')
    statements.append('f.close()')
    # Write generated codes into script file
    sfile = open(scriptfile, 'w')
    sfile.writelines(statements)
    sfile.close()
    return statements
```

ACKNOWLEDGMENTS

The author would like to thank HJ Wang (Nanyang Technological University, Singapore) for her discussion and comments on the initial drafts.

APPENDIX A
(EXAMPLE CODE ON NON-ODE / MODIFYING FUNCTION)

```
import random
import ode

birth = 0                   # birth rate
death = 0.0001              # natural death percent (per day)
transmission = 0.0095       # transmission percent (per day)
resurect = 0.0001           # resurect percent (per day)
destroy = 0.0001            # destroy percent (per day)

def human(t, y):
    infected = transmission*y[0]*y[1]
    dead = death*y[0]
    return birth - infected - dead
def zombie(t, y):
    newly_infected = transmission*y[0]*y[1]
    resurrected = resurect*y[2]
    destroyed = destroy*y[0]*y[1]
    return newly_infected + resurrected - destroyed
def dead(t, y):
    natural_death = death*y[0]
    destroyed_zombies = destroy*y[0]*y[1]
    created_zombies = resurect*y[2]
    return natural_death + destroyed_zombies - \
           created_zombies

def influx(y, step):
    y[0] = y[0] + (5*step)
    return y

f = [human, zombie, dead]   # system of ODEs
# initial human, zombie, death population, and total
y = [500.0, 0, 0]

print('Solving using 5th order Dormand-Prince \
method ......')
noinflux = [x for x in ode.DP5(f, 0.0, y, 0.1, 50.0)]
influx = [x for x in ode.DP5(f, 0.0, y, 0.1, 50.0, influx)]

for i in range(len(noinflux)):
    consolidated = noinflux[i] + influx[i][1:]
    print ','.join([str(x) for x in consolidated])
```

Appendix B
(Example Code on Boundary Values)

```
import ode

birth = 0                    # birth rate
death = 0.0001               # natural death percent (per day)
transmission = 0.0095        # transmission percent (per day)
resurect = 0.0001            # resurect percent (per day)
destroy = 0.0001             # destroy percent (per day)

def human(t, y):
    infected = transmission*y[0]*y[1]
    dead = death*y[0]
    return birth - infected - dead
def zombie(t, y):
    newly_infected = transmission*y[0]*y[1]
    resurrected = resurect*y[2]
    destroyed = destroy*y[0]*y[1]
    return newly_infected + resurrected - destroyed
def dead(t, y):
    natural_death = death*y[0]
    destroyed_zombies = destroy*y[0]*y[1]
    created_zombies = resurect*y[2]
    return natural_death + destroyed_zombies - \
           created_zombies

f = [human, zombie, dead]   # system of ODEs
y = [500.0, 0, 0]  # initial human, zombie, death population
lower_bound = {0: [10.0, 10.0]}
upper_bound = {1: [1000.0, 1000.0]}

print('Solving using 5th order Dormand-Prince \
method ......')
nobound = [x for x in ode.DP5(f, 0.0, y, 0.1, 50.0)]
lowerbound = [x for x in ode.DP5(f, 0.0, y, 0.1, 50.0, None,
                                  lower_bound)]
doublebound = [x for x in ode.DP5(f, 0.0, y, 0.1, 50.0,
                        None, lower_bound, upper_bound)]
for i in range(len(nobound)):
    consolidated = nobound[i] + lowerbound[i][1:] + \
                   doublebound[i][1:]
    print ','.join([str(x) for x in consolidated])
```

APPENDIX C
(EXAMPLE CODE ON BOUNDED VERSUS UNBOUNDED VARIABLES)

```
import ode

# ---------------------------
# Case 1: Bounded Variables
# ---------------------------

birth = 0                 # birth rate
death = 0.0001            # natural death percent (per day)
transmission = 0.0095     # transmission percent (per day)
resurrect = 0.0001        # resurect percent (per day)
destroy = 0.0001          # destroy percent (per day)

def human1(t, y):
    infected = transmission * y[0] * y[1]
    dead = death * y[0]
    return birth - infected - dead
def zombie1(t, y):
    newly_infected = transmission * \
        y[0] * y[1]
    resurrected = resurrect * y[2]
    destroyed = destroy * y[0] * y[1]
    return newly_infected + resurrected - \
           destroyed
def dead1(t, y):
    natural_death = death * y[0]
    destroyed_zombies = destroy * \
        y[0] * y[1]
    created_zombies = resurrect * y[2]
    return natural_death + destroyed_zombies - \
           created_zombies

# system of ODEs
bounded = [human1, zombie1, dead1]

# initial human, zombie, death population
y = [500.0, 0, 0]
print('Solving using 5th order Dormand-Prince method \
with bounded variables......')
for i in [x for x in ode.DP5(bounded, 0.0, y, 0.1, 50.0)]:
    print(','.join([str(z) for z in i]))

# ------------------------------
# Case 2: Un-Bounded Variables
# ------------------------------
```

```
def human2(t, y):
    infected = y[5] * y[0] * y[1]
    dead = y[4] * y[0]
    return y[3] - infected - dead
def zombie2(t, y):
    newly_infected = y[5] * y[0] * y[1]
    resurrected = y[6] * y[2]
    destroyed = y[7] * y[0] * y[1]
    return newly_infected + resurrected - \
           destroyed
def dead2(t, y):
    natural_death = y[4] * y[0]
    destroyed_zombies = y[7] * y[0] * y[1]
    created_zombies = y[6] * y[2]
    return natural_death + destroyed_zombies - \
           created_zombies

unbounded = range(8)
unbounded[0] = human2
unbounded[1] = zombie2
unbounded[2] = dead2

y = range(8)
y[0] = 500.0 # initial human population
y[1] = 0.0   # initial zombie population
y[2] = 0.0   # initial death population
y[3] = 0     # birth rate
y[4] = 0.0001 # natural death percent / day
y[5] = 0.0095 # transmission percent  / day
y[6] = 0.0001  # resurrect percent / day
y[7] = 0.0001  # destroy percent / day

print('Solving using 5th order Dormand-Prince method \
with un-bounded variables......')
for i in [x for x in
          ode.DP5(unbounded, 0.0, y, 0.1, 50.0)]:
    print(','.join([str(z) for z in i]))
```

REFERENCES

Bao, G, and Zhang, H. (2014) Sensitivity Analysis of an Inverse Problem for the Wave Equation with Caustics. *Journal of the American Mathematical Society* 27, 953-981.

Cash, JR, and Karp, AH. (1990) A Variable Order Runge-Kutta Method for Initial Value Problems with Rapidly Varying Right-Hand Sides. *ACM Transactions on Mathematical Software* 16, 201-222.

Chay, ZE, Goh, BF, and Ling, MHT. (2016) PNet: A Python Library for Petri Net Modeling and Simulation. *Advances in Computer Science: an international journal* 5(4): 24-30.

Dormand, JR, and Prince, PJ. (1980) A Family of Embedded Runge-Kutta Formulae. *Journal of Computational and Applied Mathematics* 6, 19–26.

Langtangen, HP, and Wang, L. (2014) *Odespy software package*. URL: https://github.com/hplgit/odespy.

Ling, MHT. (2016a) Of (Biological) Models and Simulations. *MOJ Proteomics & Bioinformatics* 3(4): 00093.

Ling, MHT. (2016b) COPADS IV: Fixed Time-Step ODE Solvers for a System of Equations Implemented as a Set of Python Functions. *Advances in Computer Science: an international journal* 5(3): 5-11.

Munz, P, Hudea, I, Imad, J, and Smith, RJ. (2009) When Zombies Attack!: Mathematical Modelling of Outbreak of Zombie Infection. *Infectious Disease Modelling Research Progress* 4, 133-150.

Oliphant, TE. (2007) Python for Scientific Computing. *Computing in Science & Engineering* 9, 10-20.

Otto, SP, and Day, T. (2007) *A Biologist's Guide to Mathematical Modeling in Ecology and Evolution* (Vol. 13). Princeton University Press.

Schober, M, Duvenaud, DK., and Hennig, P. (2014) Probabilistic ODE solvers with Runge-Kutta means. In *Advances in neural information processing systems*, pp. 739-747.

Date Submitted: November 9, 2017. *Date Accepted:* December 15, 2017.

About the Editor

Maurice HT Ling
Principal Partner
Colossus Technologies LLP, Singapore

School of BioSciences
The University of Melbourne, Australia

mauriceling@colossus-tech.com
mauriceling@acm.org
computer.in.science@gmail.com

Series Description

Current STEM is a broad-spectrum book series targeting practitioners, both academic and industrial practitioners, at all levels of STEM (Science, Technology, Engineering, and Mathematics). This includes all philosophical, theoretical and applied aspects of STEM; as well as STEM-related areas, such as STEM education, STEM industry and economy, ethics and legal aspects of STEM. Hence, Current STEM considers a wide variety of articles as book chapters; this includes, but not limited to:

- Original studies/works (such as research articles, and case studies)
- Updates to previous published studies/works
- Updates or additions to previous methods
- Short communications
- Null results
- Project proposals
- Reviews on research studies / methodologies / tools, etc.
- Algorithm descriptions and software codes
- Educational materials (such as tutorials, and laboratory exercises)
- Dataset communications
- Whitepapers
- Commentaries
- Opinions

In addition to book chapter submissions, Current STEM considers the following:

- Research monographs, such as, theses
- Educational monographs, such as, textbooks
- Collated conference proceedings
- Edited thematic volumes

Referee Statement

Unless otherwise specified in the book chapters, every chapter is reviewed by editorial team, supported by reviewers who are either experts or practitioners in the field, before acceptance. In event where a member of the editorial team is an author; the acceptance recommendation(s) of the external reviewer(s), whom is/are not member(s) of the editorial team, take precedence over that of the editorial team. The corresponding author will receive a review report and a letter of acceptance (if accepted) as proof of refereeing.

Book Chapter Submissions

The book chapter be attached (in RTF, DOC, DOCX, or ODF formats) in an email addressed to the Series Editor, Maurice HT Ling, at computer.in.science@gmail.com, with a cover letter stating (1) a summary of the current submission and its significance, and (2) contact details of 3 or more potential reviewers (the author may also list unsuitable reviewers).

Monographs and Proceedings

Research or educational monographs, and collated conference proceedings submissions should be discussed with the Series Editor,

Maurice HT Ling, via email correspondence at computer.in.science@gmail.com, prior to submission as a book proposal is needed for each volume.

EDITED THEMATIC VOLUMES

Current STEM invites experts to serve as Guest Editors for individual edited thematic volumes. An edited thematic volume collection of scholarly or scientific chapters written by different authors collected together in one book by Guest Editor(s). The chapters in an edited volume must be original works (not republished works). As each volume requires a book proposal, potential Guest Editor(s) discuss their intent with the Series Editor, Maurice HT Ling, via email correspondence at computer.in.science@gmail.com.

INDEX

A

Advanced Encryption Standard (AES), 131, 139, 140, 141
age, 13, 16, 25, 43, 51, 72, 73, 75, 76, 78, 109, 111, 112, 118, 119, 124
AIDS, 108, 109, 110, 125, 126, 127, 128
algorithm, 3, 11, 67, 68, 72, 77, 84, 85, 104, 131
algorithmic beauty, 144, 166, 168, 170
anticoagulant, 2, 105
application programming interface, 5, 14
artificial life, 1, 2, 3, 10, 11, 32, 33, 103, 104
automata, 3, 104

B

Babington Plot, 130, 132
bacteria, 2, 10, 24
bacterium, 10
Barbados, 117, 119, 126, 127
behaviors, 108, 109, 111, 118, 119, 120, 121, 122, 123, 124, 126
being state transition modelling, 174
bias, 118, 123
biology, 1, 2, 3, 11, 67, 68, 72, 77, 84, 85, 103, 104, 105, 169, 170, 171
birth rate, 183, 207, 208, 209, 210
Bletchley Park, 132
blood, 5, 22, 23, 29, 43, 51, 62, 63, 72, 73, 145, 169
blood vessels, 145
BlowFish, 131, 140
branching, 143, 144, 169
bronchial tree, 169
Butcher Tableaues, 177

C

C++, 145
calculus, 178, 185
carbon, 10, 11
Caribbean, 108, 109, 110, 120, 125, 126, 127
case studies, 4, 215
Cash-Karp method, 177, 190, 192

cellular automata, 3, 104
challenges, 107, 110
chemical reactions, 3, 10
chromosome, 5, 13, 15, 22, 23, 29, 30, 31, 32, 38, 39, 44, 45, 46, 47, 51, 52, 56, 62, 63, 68, 70, 71, 72, 73, 74, 75, 76, 82, 83, 84, 86, 87, 90, 94, 95, 98, 100, 101
clean air, 104
clone, 76, 84
code generation, 149
coding, 144
codon, 12, 30, 46, 47, 63, 87, 90, 95, 98, 101
college students, 118
comparative analysis, 141
competing interests, 124
complexity, 133, 134, 135, 140
complications, 108
computer, 10, 144, 213, 216, 217
computer simulations, 11
computing, 131
cross-sectional study, 107, 111
cryptography, v, 129, 130, 131, 132, 133, 141

D

data analysis, 124
data collection, 125
data structure, 175
database, 1, 3, 4, 5, 6, 7, 9, 13, 14, 20, 22, 23, 27, 28, 30, 31, 32, 36, 37, 38, 39, 40, 41, 42, 43, 44, 45, 46, 48, 50, 52, 53, 54, 56, 67, 87, 89, 91, 93, 95, 97, 98, 99, 00, 102, 103
deaths, 175, 180
decoding, 130
deficiency, 129
demographic characteristics, 113
demographic factors, 108
demonstrations, 165
dependent variable, 110
differential equations, 173, 178, 185
diffusion, 19
digital organism, v, 1, 2, 3, 9, 11, 12, 14, 32, 103, 104
DNA, 2, 3, 12
DOC, 216
Dormand-Prince method, 177, 197, 199, 209, 210
DOSE, v, 1, 3, 4, 5, 6, 9, 11, 12, 13, 14, 32, 53, 86, 87, 90, 94, 101, 103, 104
drawing, 151

E

ecological systems, 125
ecology, 11, 18, 19, 31, 33, 35, 56, 88, 92, 96, 99, 102
economic status, 108, 109, 117, 123
economically motivated relationships, 108, 121, 126
ecosystem, 4, 5, 6, 12, 13, 17, 18, 19, 20, 21, 23, 24, 25, 27, 30, 31, 33, 34, 35, 36, 41, 42, 43, 49, 59, 60, 61, 62, 63, 72, 91, 95, 96
education, 109, 118, 123, 215
educational background, 118
emergent features, 3
encryption, 129, 130, 131, 132, 133, 136, 137, 139, 140, 141
encryption for transportation, 130
environment, 2, 9, 14, 32, 118, 123
environmental resources, 62
environments
enzyme, 179, 180
epidemic, 16, 108, 109, 126
epydoc, 175
Euler method, 177, 184, 186, 187, 188, 189, 191, 192, 194, 196, 197, 199
evolution, 1, 2, 3, 11, 103, 104, 105, 211

execution, 9, 37, 62, 145, 146, 147, 148, 206
experimental evolution, 2, 3, 103
exposure, 118, 124
extraction, 2
extracts, 42, 43, 44, 45

F

family relationships, 111
file splitting, 130
first generation, 9
fitness, 15, 30, 31, 43, 51, 63, 64, 72, 73, 74, 77, 78, 79, 82, 83, 88, 92, 96, 99, 102
fixed time step, 174
force, 30, 129, 131, 133

G

genetic algorithm, 3, 11, 32, 33, 67, 68, 72, 77, 84, 85, 104
genetic analysis, 2
genetic code, 12, 38, 104
genome, 2, 3, 5, 10, 12, 13, 15, 22, 23, 29, 30, 31, 44, 45, 46, 47, 50, 52, 62, 63, 68, 69, 70, 72, 73, 74, 75, 76, 78, 82, 83, 84, 86, 87, 88, 90, 92, 94, 95, 96, 99, 100, 101, 102, 104
genre, 108, 126
geometry, 166
greenhouse, 171
Grenada, v, 107, 109, 110, 112, 117, 125, 127

H

health, 108, 110, 111, 115, 118, 120, 123, 126
health risk behaviors, 108, 109, 111, 118, 123
health risks, 123
heterozygosity, 4
HIV, v, 107, 108, 109, 112, 117, 124, 125, 126
HIV/AIDS, 107, 108, 109, 110, 125, 126
hotspots, 68
human, 17, 20, 131, 169, 174, 175, 176, 180, 181, 182, 183, 203, 204, 207, 208, 209, 210
hybridization, 105

I

identity, 5, 16, 22, 23, 26, 43, 44, 45, 50, 58, 71, 72, 73, 74, 89, 92, 93, 97, 102
income, 108, 123
independent variable, 108
individuals, 123
information exchange, 130
information processing, 211
initiation, 117, 118
integration, 178, 186, 204
interface, 5
intimate partner violence, 108, 109, 122, 124, 128
inversion, 68, 70, 134
iteration, 143, 144

J

jigsaw, v, 129, 130, 132, 133, 134, 135, 136, 137, 138, 139, 140
jigsaw encryption system, 129, 132
jigsaw puzzle, v, 129, 132

L

Langton, Christopher, 3
Lindenmayer system, v, 143, 144, 146, 169, 170

Lindenmayer, Aristid, 144
logging, 1, 3, 5, 9, 20, 30, 31, 32, 36, 37, 38, 39, 40, 46, 48, 53, 54, 56, 87, 91, 95, 98, 99, 102
L-system, 143, 144, 145, 146, 169, 170, 171

M

mapping, 5, 11, 149, 150, 166, 167
mathematical modelling, 173, 174, 211
mathematics, vii
measurement, 63, 64
medical science, 126
meiosis, 15
memory, 76, 77, 79
mental health, 126
migration, 4, 86, 87, 89, 90, 93, 94, 95, 98, 104, 105
modelling, 173, 174
modifying function, 174, 207
morphogenesis, 169
morphology, 171
multicellular organisms, 144, 170
music, 108, 111, 120, 121
mutation, 5, 14, 16, 29, 31, 46, 47, 51, 52, 64, 68, 69, 70, 71, 72, 73, 74, 75, 80, 81, 82, 83, 84, 86, 87, 88, 90, 92, 94, 96, 97, 98, 99, 100, 101, 102
mutation rate, 29, 68, 73, 75, 81, 83, 86, 90, 94

N

native species, 105
negotiation, 109, 117, 127

O

ODE solver, v, 173, 174, 176, 177, 179, 184, 204, 211
operations, 3, 14, 53, 147
ordinarily differential equation, 174
organism, 2, 3, 5, 6, 9, 11, 12, 14, 15, 17, 18, 20, 21, 22, 30, 31, 32, 33, 35, 56, 58, 61, 62, 64, 71, 72, 73, 74, 75, 76, 77, 78, 79, 80, 83, 86, 87, 88, 89, 90, 91, 92, 93, 94, 95, 96, 97, 99, 100, 102

P

parametric-stochastic system, 144
parents, 5, 16, 23, 43, 51, 72, 73, 88, 89, 92, 96, 97, 99, 100, 114
participants, 111, 112, 113, 117, 118, 119, 120, 122, 123
peppered moth, 2, 104
petroleum, 170
phenotype, 2
point mutation, 73, 86, 90, 94
population, 3, 5, 6, 11, 13, 14, 15, 16, 17, 18, 20, 21, 22, 24, 25, 26, 27, 28, 29, 30, 31, 38, 39, 41, 43, 44, 45, 46, 47, 50, 52, 53, 54, 55, 58, 59, 60, 61, 62, 63, 64, 65, 66, 67, 72, 73, 76, 77, 78, 79, 80, 81, 82, 83, 84, 85, 86, 87, 88, 89, 90, 91, 92, 93, 94, 95, 96, 97, 98, 99, 100, 101, 102, 109, 110, 119, 124, 181, 183, 207, 208, 209, 210
population control, 15, 16, 31, 77, 78, 79, 83, 84
population group, 110
population size, 28, 77, 80
poverty, 117, 121
probability, 68, 73, 75, 90, 94, 145, 146, 147, 148
programming, 5, 135
public health, 125, 126
P-value, 113, 114
python, 1, 3, 6, 9, 10, 11, 12, 21, 22, 23, 24, 32, 33, 67, 68, 72, 77, 84, 85, 86, 87, 90, 94, 101, 104, 129, 130, 135, 143, 144,

145, 146, 148, 173, 174, 176, 178, 179, 180, 182, 204, 211
python module, 10, 144

Q

query, 40, 41, 49, 50
questionnaire, 108, 111, 119

R

replacement rules, 145, 147, 148
resistance, 2, 3, 4, 103, 104, 105, 122
resources, 19, 62, 108, 124
rewriting rules, 143, 144, 146
risk, 107, 109, 111, 115, 117, 118, 119, 120, 121, 123, 124, 125, 126
risk factors, 107, 109, 117, 120, 124, 125
rodenticide, 2
Runge-Kutta method, 177, 187, 188, 189, 211
Runge-Kutta-Fehlberg method, 177

S

SciPy, 174, 176
script generator, v, 173, 176, 179, 180
scripts, 180
secondary school education, 118
security, 129, 139, 141
self-documenting code, 175
self-employed, 112
sensor network, 141
sexual abuse, 122
sexual behavior, 120
sexual health, 109
sexual intercourse, 109, 111, 118, 119
sexual risk behaviors, 108, 119, 121
sexual violence, 109
significance level, 108
simulation, v, 1, 2, 3, 4, 5, 6, 7, 8, 9, 11, 12, 13, 14, 17, 20, 21, 22, 23, 24, 27, 28, 29, 30, 31, 32, 33, 37, 38, 39, 40, 41, 42, 43, 44, 45, 46, 47, 48, 49, 50, 52, 53, 54, 55, 56, 57, 58, 59, 60, 61, 62, 64, 65, 66, 67, 86, 87, 89, 90, 91, 93, 94, 95, 96, 97, 98, 99, 101, 102, 103, 104, 169, 170, 179, 183, 202, 204, 211
simulation environment, v, 1, 2, 3, 9, 11, 12, 14, 32, 103, 104
Singapore, 1, 129, 143, 173, 206
social construct, 123, 124, 128
socio-demographic factors, 108
software, 211, 215
South Africa, 122, 126, 127
speciation, 3, 86, 89, 93, 103
spectrum management, 18, 31
sphinx, 175
storage, 53, 129, 139, 140
string rewriting, 143, 144, 145

T

Tanzania, 125, 127
transformations, 22, 38, 129
translocation, 68, 70
transmission, 133, 174, 175, 176, 180, 181, 183, 203, 207, 208, 209, 210
transportation, 129, 130, 139, 140
turtle, 144, 146, 166, 167
turtle graphics, 148, 166, 168

V

variables, 111, 178, 179, 184, 185, 201, 202, 204, 205, 209, 210
variations, 134
vector, 149, 173, 176, 178, 179
violence, 108, 109, 122, 124, 126, 128

W

warfarin, 2
West Indies, 127
World Health Organization (WHO), 108, 125
World War I, 132

Y

yield, 1, 177, 178, 186, 187, 188, 189, 190, 191, 192, 194, 195, 196, 197, 199, 201